A New Guide
to Modern
Valency Theory

OTHER BOOKS BY THE SAME AUTHOR

Introduction to Chemistry
Introduction to Organic Chemistry
Introduction to Physical Chemistry

A New Guide to Modern Valency Theory

SI Edition

G. I. Brown
Assistant Master, Eton College

With a foreword by
L. E. Sutton M.A. D.PHIL. F.R.S.

LONGMAN

LONGMAN GROUP LIMITED
London
Associated companies, branches and representatives throughout the world

© Longman Group Ltd 1967–1972

All rights reserved. No part of this publication may be reproduced, stored in a retrieval system or transmitted in any form or by any means, electronic, mechanical, photocopying, recording or otherwise, without the prior permission of the copyright owner.

First published 1953
Tenth impression 1964
Second edition 1967
Third impression 1970
SI edition 1972
Third impression 1979

ISBN 0 582 35024 7

Printed in Great Britain by
Lowe & Brydone Printers Limited, Thetford, Norfolk

Foreword

One of the most agreeable of a College tutor's duties is to keep an eye on the activities of his former pupils. Sometimes this involves him in hard work; and so it is in the present case for Mr. G. I. Brown asked me to look critically at the first edition of his book on valency and again at this, the second, edition.

Since the first edition was published there certainly has been one major extension of ideas, namely the ligand field theory of complexes. By chance, another old pupil—Dr. L. E. Orgel, F.R.S.—played a leading part in its development. There have also been many applications which, though individually of a more detailed nature, represent in total a very considerable advance in understanding. There has, too, been a great deal of work aimed at making the theory more accurate and quantitative, though most of it must be deemed beyond the scope of this book. One change which is particularly notable is a big shift of emphasis from valence bond to molecular orbital treatments. This has happened partly because the latter are simpler in some respects but a major reason, I think, is that molecular orbital theory introduces symmetry in a very illuminating way and symmetry is the branch of mathematics in which chemists most readily feel at home. The resulting oversimplifications need not immediately worry the readers of this book. Molecular orbital theory is fashionable as well as useful, and they must know about it. Altogether, therefore, there has been a lot that is new to write about and to present in a simple way. I think that Mr. Brown has done a difficult task well.

The fundamental concepts of wave mechanics still remain subtle and difficult. There is a danger that a new dogma will develop, based on theory imperfectly understood and used too confidently and uncritically. It is unlikely that readers of this book will fall too deeply into this particular form of sin; but, if they find this introduction to the subject as interesting as we hope they will and want to know more, they would be well advised to study the subject again from the beginning, and to do so more rigorously, so that they can apply it and gain the great benefits that it can give, while appreciating its limitations and difficulties. All learning is a series of approximations to truth. This book provides a very reasonable and readable first approximation to a particular piece of truth.

L. E. SUTTON

April 1967
Magdalen College
Oxford

Note to SI Edition
The units used in this edition have been changed into SI units, and the names of chemicals have been brought up-to-date.

Contents

Foreword		v
Preface		ix
1	Introduction	1
2	Quantum Numbers	7
3	The Periodic Table	20
4	Types of Chemical Bond	32
5	The Formation of Ionic Bonds	38
6	Characteristics of Ionic Compounds	51
7	The Wave Nature of the Electron	67
8	Molecular Orbitals	81
9	Simple Examples of Covalent Bonding	91
10	Directional Nature of Covalent Bonds	100
11	Resonance and Electronegativity	117
12	Characteristics of Covalent Compounds	130
13	Hydrogen Bonding	141
14	van der Waals' Forces	154
15	Complex or Co-ordination Compounds	162
16	Crystal Field Theory	180
17	Some Applications of Crystal Field Theory	190
18	Ligand Field Theory	198
19	Bonding in Metals	209

Appendix: Radii and Energy Levels of Stationary States in the Hydrogen Atom — 228
Formula Index — 231
Subject Index — 234
Periodic Tables — 240
The Arrangement of Electrons in the Atoms of the Elements in their Normal States — 242

Preface

It is now almost twenty years since the first edition of this book was published, and the preparation of a second edition involved a complete re-writing. A much greater emphasis has been put on molecular orbital theory, and on the part played by d electrons in crystal field and ligand field theory. There is also a fuller treatment of metallic bonding and of van der Waals' bonding. The third edition brings nomenclature up-to-date, and introduces SI units.

The ideas from the first edition which still remain valid have been retained, and the basic aim of the book has not been changed. It is intended to bridge the gap which the author believes to exist between the treatment of valency theory given in standard text-books of chemistry and that given in the more advanced works devoted entirely to valency.

The book is intended for use by advanced sixth form pupils, by first-year university students, and by older chemists who are interested in modern developments but who may not have the time or inclination to study the more advanced works.

The treatment of the material is simple, concise and mainly qualitative; mathematical considerations and experimental details have been reduced to a minimum. The aim has been to give a broad, general account of modern valency theory which can be easily and readily grasped, and which will, it is hoped, help towards a fuller understanding of chemistry as a whole.

1 Introduction

1 The importance of valency A study of chemistry very soon reveals the wide variety and immense number of different chemical compounds. At an elementary level the learning of the subject can degenerate into an attempt to learn a seemingly endless catalogue of unrelated facts. In general, many students find too few threads to which they can cling, and, as a result, it is difficult to obtain any real understanding.

Great progress has been made in recent years towards gaining a greater insight into the nature of chemical reactions and chemical bonding between atoms. In the past, the treatment has been largely experimental. It has been found, for instance, that substance A combined with substance B to form substance C, and the conditions for such a reaction to take place have been worked out in some detail, as have the particular properties of A, B and C. But the more fundamental questions as to why, or at least how, A and B combine to form C, why the properties of A, B and C are what they are, and why X and Y, for instance, will not combine, were, up till recent times, left unanswered if not unasked.

The study of the problems which such questions raise is now called the study of valency. The word originates from the late Latin, valentia, meaning strength, and it is with a study of the combining powers of chemical substances and the nature of chemical bonds that valency is largely concerned.

2 The dualistic theory The earliest theory of valency which had any success in explaining the known facts was put forward by Berzelius in 1812. Following, as it did, the discovery of the Voltaic pile in 1800 and the newly discovered results of early experiments on electrolysis, it is not surprising that the theory was of an electrical nature.

In electrolysis different substances appear at the two electrodes and Berzelius allotted to the atoms of each element an electrical polarity so that some atoms had a positive charge and some a negative one. Chemical combination, he said, took place between atoms with different charges as, for instance, in the formation of sodium chloride from positively charged sodium atoms and negatively charged chlorine atoms:

$$Na^+ + Cl^- \rightarrow NaCl$$

To account for the building up of larger molecules it was only necessary to assume an incomplete neutralisation of charge so that the formation of hydrated iron(II) sulphate(VI), for instance, was envisaged according to the scheme:

$$Fe^+ + O^- \rightarrow FeO^+$$
$$S^+ + 3O^- \rightarrow SO_3^-$$
$$FeO^+ + SO_3^- \rightarrow FeSO_4^-$$
$$2H^+ + O^- \rightarrow H_2O^+$$
$$FeSO_4^- + 7H_2O^+ \rightarrow FeSO_4.7H_2O$$

Atoms combine, on this theory, because they attract each other electrically. This idea is still an essential part of modern valency theory at least so far as ionic substances are concerned (page 32).

3 The theory of types The dualistic theory is clearly capable of considerable extension using the idea of residual electrical charges remaining on a compound after formation from its component atoms. As organic chemistry developed in the middle and later part of the nineteenth century, however, Berzelius' theory was replaced by a theory of types due originally to Dumas in 1839.

The dualistic theory failed to explain the fact that a supposedly positive atom could be replaced in a chemical compound by a supposedly negative atom without any great change in the nature of the compound. Thus ethanoic acid and chloroethanoic acid, or manganic(VII) and chloric(VII) acids are not dissimilar, though on the dualistic theory, a hydrogen or a manganese atom is quite distinct from a chlorine atom.

CH_3COOH — Ethanoic acid
$CH_2ClCOOH$ — Chloroethanoic acid
$HMnO_4$ — Manganic(VII) acid
$HClO_4$ — Chloric(VII) acid

Furthermore, with the acceptance of Avogadro's hypothesis, it became clear that the molecules of many elements are polyatomic, e.g. H_2, O_2, Cl_2, so a theory was needed which could account for the linking together of two like atoms.

Dumas' idea was that there were certain fundamental types of chemical compound and that any element or group of elements in these types could be replaced, equivalent for equivalent, by another element or group of elements. This theory was developed by Williamson and Gerhardt, and the latter propounded four types, allotting known compounds to each type as illustrated below:

Hydrogen type
$\left.\begin{array}{c}H\\H\end{array}\right\}$

Hydrochloric acid type
$\left.\begin{array}{c}H\\Cl\end{array}\right\}$

Water type
$\left.\begin{array}{c}H\\H\end{array}\right\}O$

Ammonia type
$\left.\begin{array}{c}H\\H\\H\end{array}\right\}N$

Hydrocarbons, e.g.
$\left. \begin{array}{c} CH_3 \\ H \end{array} \right\}$

Alkyl halides, e.g.
$\left. \begin{array}{c} CH_3 \\ Cl \\ C_2H_5 \\ Br \end{array} \right\}$

Alcohols, e.g.
$\left. \begin{array}{c} CH_3 \\ H \end{array} \right\} O$

Amines, e.g.
$\left. \begin{array}{c} CH_3 \\ H \\ H \end{array} \right\} N$

Aldehydes, e.g.
$\left. \begin{array}{c} C_2H_3O \\ H \end{array} \right\}$

Ethers, e.g.
$\left. \begin{array}{c} CH_3 \\ CH_3 \end{array} \right\} O$

Amides, e.g.
$\left. \begin{array}{c} C_2H_3O \\ H \\ H \end{array} \right\} N$

Ketones, e.g.
$\left. \begin{array}{c} C_2H_3O \\ CH_3 \end{array} \right\}$

Carboxylic acids,
$\left. \begin{array}{c} C_2H_3O \\ H \end{array} \right\} O$

Once again the theory is clearly capable of including a large number of known compounds, particularly when condensed types (Williamson), mixed types (Kekulé), and further simple types are introduced.

The theory was very successful in the realm of organic chemistry and eventually developed into the idea of *homologous series*. The main emphasis was on the structure of the molecule. The dualistic theory was concerned more with the nature of the particles combining than with their arrangement within a molecule.

A long and bitter controversy ensued between the exponents of the two rival theories, and, though the dualistic theory came into its own again when electrical ideas came to the fore as a result of Arrhenius' work in and after 1887, it is now realized that the two theories were dealing with different kinds of compound. In general the ideas of Berzelius apply in a modified form to what we now call electrolytes, whereas Dumas was dealing with non-electrolytes. In judging the merits of these old theories it is important to remember that they were developed at a time when atomic weights were not known with any accuracy and when it was far from certain that the formula of water was H_2O.

4 Valency as a number As the number of known chemical compounds grew, and as their formulae became known with more accuracy, the similarity of many of them became apparent. This was first noticed by Frankland, a supporter of Berzelius, in 1852, particularly in such compounds of nitrogen, phosphorus, and arsenic as*

NH_3		N_2O_3	N_2O_5	NH_4Cl
PH_3	PCl_3	P_2O_3	P_2O_5	PH_4Cl
AsH_3	$AsCl_3$	As_2O_3		

in which the three elements are always combined with either three or five atoms of some other elements.

---

* The formulae given are the modern ones; Frankland used different formulae for some of the compounds but the similarity was still clear.

Frankland put forward the suggestion that an atom of an element had a certain 'combining power' which determined the number of atoms of another element with which it would combine. Thus nitrogen, phosphorus and arsenic require three or five atoms of some other element to satisfy their 'combining power'.

This early suggestion led to the idea of the valency of an element being expressed as a number which gave a quantitative measure of the 'combining power' of the element. Moreover, the choice of hydrogen as the unit of 'combining power' made possible *the definition of the valency of an element as the number of atoms of hydrogen with which one atom of the element would combine*. The development of the relationship

$$\text{Relative atomic mass } (A_r) = \text{equivalent mass} \times \text{valency}$$

stressed this idea of valency as a number.

There followed much argument as to whether the valency of an element was fixed or variable, but with the acceptance of the latter view in certain cases the simple statement became, and still is, a useful definition. It will be found in nearly all elementary chemistry books and is normally a schoolboy's first introduction to the word valency. By allotting the correct valency number (or numbers) to each atom or radical it is possible to build up the correct formulae for many chemical compounds. Thus, in general, if the formula of a compound is $A_x B_y$, A and B being elements or radicals, then

$$x \times \text{Valency of } A = y \times \text{Valency of } B.$$

A knowledge of the correct numerical valencies of elements and radicals is, therefore, important in writing correct chemical formulae, and valency tables can be drawn up to serve as useful mnemonics.

A numerical valency value gives, however, no information as to the nature of the chemical bonds formed between atoms. The whole idea of valency as a number must, in fact, be treated with some caution as explained on page 127.

Modern valency theory goes far beyond the simple idea of valency as a number and attempts to account for the detailed mechanism of formation of chemical bonds and their detailed nature once they are formed. This is done by making use of the facts and ideas which came to light in the development of the study of atomic structures.

5 Summary of results of work on atomic structure

The full story of the work leading to the elucidation of atomic structure is both long and complex. All that is required, here, is a concise summary of the major results.

(a) Atoms are made up of three fundamental particles, which differ in mass and electrical charge as follows:

	Comparative mass	Comparative charge	Actual rest mass/kg	Actual charge/C
Electron	1/1837 unit	−1 unit	$9{\cdot}109\ 08 \times 10^{-31}$	$1{\cdot}602\ 10 \times 10^{-19}$
Proton	1 unit	+1 unit	$1{\cdot}672\ 52 \times 10^{-27}$	$1{\cdot}602\ 10 \times 10^{-19}$
Neutron	1 unit	Not charged	$1{\cdot}674\ 82 \times 10^{-27}$	Not charged

The masses quoted for the three particles are the rest masses. When the particles are moving their mass increases according to relativity theory.

(b) An atom consists of a heavy, positively charged, central nucleus containing protons and, excepting the hydrogen atom, neutrons, around which electrons are distributed. The positive charge on the nucleus, caused by the protons, is neutralised by the negative charge of the electrons so that the atom as a whole is electrically neutral. To achieve this the number of protons and electrons in an atom must be equal.

(c) The mass of an atom is made up, almost entirely, of the mass of the protons and neutrons.

(d) In passing from atom to atom in the periodic table there is a unit increase in positive charge on the nucleus, and in the number of electrons. The *atomic number* (Z) of an element represents its ordinal number in the periodic table (e.g. hydrogen, 1; uranium, 92), the positive charge on the nucleus of the atom, and the number of electrons in the atom.

Typical simple atomic structures are shown as follows:

Hydrogen $A_r = 1$, $Z = 1$ — (1p) 1e

Helium $A_r = 4$, $Z = 2$ — (2p 2n) 2e

Lithium $A_r = 7$, $Z = 3$ — (3p 4n) 3e

Beryllium $A_r = 9$, $Z = 4$ — (4p 5n) 4e

Carbon $A_r = 12$, $Z = 6$ — (6p 6n) 6e

Nitrogen $A_r = 14$, $Z = 7$ — (7p 7n) 7e

(e) The chemical properties of an atom, as will be seen, depend on the arrangement of the electrons in the atom, and atoms may have the same number and arrangement of electrons but different numbers of neutrons in the nucleus. This possibility leads to the existence of *isotopes* — atoms with the same chemical properties but different relative atomic masses. The structures of the three isotopes of hydrogen and of uranium are shown below:

Hydrogen, or protium, ^1_1H	Heavy hydrogen, or deuterium, ^2_1H or D	Tritium, ^3_1H or T
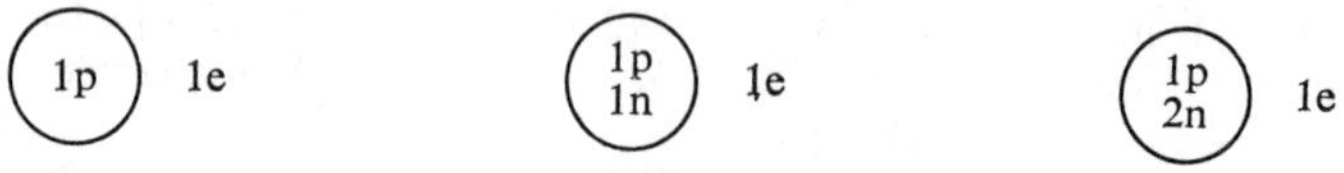 (1p) 1e	(1p, 1n) 1e	(1p, 2n) 1e
Uranium-238, $^{238}_{92}U$	Uranium-235, $^{235}_{92}U$	Uranium-234, $^{234}_{92}U$
 (92p, 146n) 92e	(92p, 143n) 92e	(92p, 142n) 92e

(*f*) An understanding of the chemical properties of an atom depends on an understanding of the way in which the electrons are arranged in that atom, and the arrangement of electrons in atoms is the theme of the next two chapters.

In developing the theme it is necessary to make use of experimental spectroscopic results and the ideas of the quantum theory. It is also necessary to take into account the fact that electrons have both a particle-like and a wave-like character, the latter being demonstrated by the fact that a beam of moving electrons can be diffracted (p. 68).

2 Quantum Numbers

The arrangement of the electrons in an atom is of quite fundamental importance because the chemical properties of an atom are largely controlled by it. This is because it is the interaction between the outer electrons of two or more atoms that leads to chemical combination. The working out of the arrangement of electrons in an atom is an amazing feat similar to the fitting together of a most complicated jig-saw puzzle or the solution of an elaborate cipher. Many of the experimental clues come from the study of atomic spectra and are interpreted by applying the ideas of the quantum theory. So-called quantum numbers are used for labelling different electrons in an atom.

SPECTRA
1 Types of spectra As opposed to particle-like 'rays', wave-like radiation can cause interference patterns, can be diffracted, if a suitable diffraction grating can be found (page 23), and can be

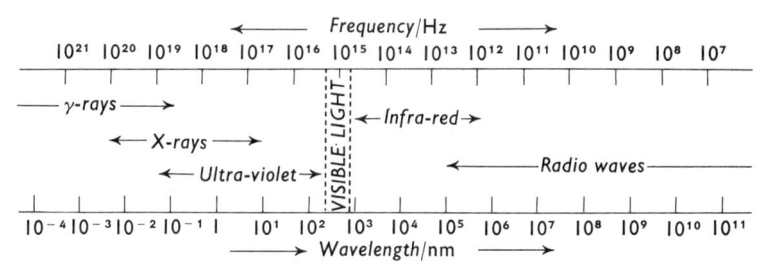

⟵ Frequency/Hz ⟶
10^{21} 10^{20} 10^{19} 10^{18} 10^{17} 10^{16} 10^{15} 10^{14} 10^{13} 10^{12} 10^{11} 10^{10} 10^{9} 10^{8} 10^{7}

⟶ γ-rays ⟶
⟵ X-rays ⟶
⟵ Ultra-violet ⟶
VISIBLE LIGHT
⟵ Infra-red ⟶
⟵ Radio waves ⟶

10^{-4} 10^{-3} 10^{-2} 10^{-1} 1 10^{1} 10^{2} 10^{3} 10^{4} 10^{5} 10^{6} 10^{7} 10^{8} 10^{9} 10^{10} 10^{11}
⟶ Wavelength/nm ⟶

Fig. 1. The electro-magnetic spectrum. The boundaries between one type of radiation and another are not very sharply defined

'sorted out' into its component wavelengths by using a spectrometer.

In this way, visible light is readily shown to be made up of different coloured lights, each colour corresponding to a group of waves of different wavelengths. Blue light, at one end of the visible spectrum, has a shorter wavelength* (0·45 μm or 4500 Å) than red light (0·75 μm or 7500 Å), at the other end. The Ångstrom unit is the unit in which wavelengths are usually expressed; it is equal to 10^{-10} m. Visible light represents only a small part of all radiation, and a more complete representation is given in Fig. 1.

* Wavelengths in, or near to, the visible region are conveniently expressed in micrometres (μm) or nanometres (nm); 1 nm is 10^{-9} m, and 1 μm is 10^{-6} m. The older unit used for this purpose was the Ångstrom, 1 Å being equal to 10^{-10} m. Frequencies are expressed in vibrations for second, s^{-1}, or hertz (Hz).

Characteristic spectra can be obtained from substances by causing them to emit radiation. This can be done by heating a substance or by subjecting it to electrical stimulation or excitation by using an electric arc, spark or discharge. A variety of so-called *emission spectra* can be obtained, of three main types.

(*a*) *Continuous spectra.* Continuous spectra show the presence of radiation of all wavelengths over a wide range. Such spectra are given by incandescent solids, e.g. the filament in an electric light bulb.

(*b*) *Band spectra.* Band spectra consist of a series of bands of overlapping lines. They are formed by the radiation emitted from excited molecules.

(*c*) *Line spectra* (Fig. 2). These consist of a series of sharply defined lines each corresponding to a definite wavelength. They are given when the atoms in a substance are excited so that they can emit radiation. The light from a mercury vapour lamp or from solid sodium chloride heated in a Bunsen flame provides a line spectrum.

Because line spectra are caused by energy changes taking place within an atom they may also be called *atomic spectra*. They reveal the energy changes which can take place within an atom and thereby provide information which leads to the elucidation of the arrangement of electrons in the atom.

Absorption spectra are formed when light of particular wavelengths is absorbed by passage through a substance. Black lines are found in the spectrum corresponding to the wavelength absorbed.

2 Spectral series So far as the historical development of atomic structure is concerned a study of the line spectrum of hydrogen is of great importance. The spectrum is obtained by passing an electric discharge through hydrogen at a low pressure and investigating the emitted radiation in a spectrometer. There are not very many lines in the visible region of the spectrum though the spectrum as a whole contains many lines (Fig. 2). Over a period of time, starting in 1885, it was found that these lines could be fitted into series and this suggested that the lines were probably related to each other in some way.

The series, known after their discoverers as the Balmer (1885) Paschen (1896), Lyman (1915), Brackett (1922) and Pfund (1925) series, consist of lines representing wavelengths which can be expressed in one overall formula:

$$\sigma = \frac{1}{\lambda} = R_H \left(\frac{1}{n^2} - \frac{1}{m^2} \right)$$

where λ is the wavelength, σ the wave number, R_H a constant, known as Rydberg's constant, and *n* and *m* have integral values as shown in the following summary

Series	n	m	Main spectral region
Lyman	1	2, 3, 4, etc.	Ultra-violet
Balmer	2	3, 4, 5, etc.	Visible
Paschen	3	4, 5, 6, etc.	Infra-red
Brackett	4	5, 6, 7, etc.	Infra-red
Pfund	5	6, 7, 8, etc.	Infra-red

It is a remarkable experimental fact that so many apparently, at first sight, unrelated lines in a spectrum can be expressed by a single, simple formula. In fact, the wave number of any line in the hydrogen

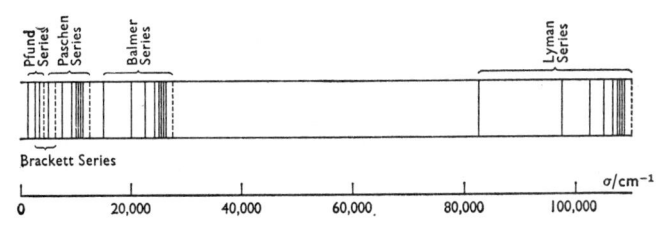

Fig. 2. The main lines in the hydrogen spectrum. The intensity of the lines varies, and the limit to each series is shown by a dotted line

spectrum can be expressed as the difference of two terms given by R_H/n^2 when *n* has two different integral values. Such values of R_H/n^2 are known as *spectral terms*.

The line spectra of the alkali metals are made up of series of lines similar to those in the hydrogen spectrum. At least four series can be detected and they are known as the *sharp, principal, diffuse* and *fundamental* series. There are more lines in the spectra of the alkali metals than in the hydrogen spectrum, and they overlap to some extent so that the existence of the series is not so readily observable. Nevertheless, all the lines can be related in a single formula much like that for the hydrogen spectrum (but with *n* and *m* not having whole-number values) or expressed as the difference of two terms.

The line spectra of all other elements also consist of series of lines but they may not be readily recognisable because of the large number of lines involved and the degree of overlapping between one series and the other.

Suggestions, by Bohr in 1914, that spectral lines originated from energy changes taking place within atoms led to the working out of the arrangement of electrons in different energy levels within atoms. To do this, Bohr applied the ideas of the quantum theory, first put forward by Planck in 1900, to the interpretation of the available spectroscopic data.

QUANTUM THEORY

3 Origin The quantum theory originated in a study of the radiation emitted by a so-called black body. This is a theoretical concept, which cannot be fully achieved in practice. An ideal black body would be a perfect absorber of all radiation, and, equally, a perfect, complete radiator. Real bodies are not perfect absorbers or radiators. Brightly polished surfaces absorb very little radiation; glass absorbs most ultra-violet light but not much visible light; a sheet of metal with a dull, black surface absorbs both ultra-violet and visible light, but may not absorb X-rays. In practice, the nearest approach to ideal black body radiation is provided by the radiation emitted through a small hole in a hot, closed furnace. The furnace can be maintained at a particular temperature and the distribution of the energy of the emitted radiation amongst different wavelengths can be investigated.

Classical theory led to the conclusions that black body radiation should consist, mainly, of wavelengths in the violet and ultra-violet regions of the spectrum, and that the intensity of the radiation should rise steadily as the wavelength fell. But this was not in agreement with the experimental facts. Lummer and Pringsheim showed, at the end of the nineteenth century, that the intensity of black body radiation, at any particular temperature, rose to a maximum and then fell again when plotted against wavelength. The maximum is in the infra-red region at low temperatures, which is why a warm, black-iron rod is invisible in the dark. As the temperature is raised, the maximum moves over into the visible region of the spectrum. The iron rod appears feebly red hot, at first, then orange-, yellow- and, finally, white-hot, as the maximum intensity of radiation moves across the visible region of the spectrum.

The failure of classical theory to account for the variation of the intensity of radiation from a black body with wavelength triggered off the quantum theory.

4 Outline of theory The essential idea of the quantum theory is that the energy of a body can only change by some definite whole-number multiple of a unit of energy known as a quantum. This means that the energy of a body cannot change continuously. It can only increase or decrease by 1, 2, 3, 4, ..., n quanta, but never by $1\frac{1}{2}$, $2\frac{3}{4}$, 107·3, etc., quanta. It is rather like the fact that our currency can only change by 1, 2, 3, 4,...,n halfpennies, but not by $1\frac{1}{2}$, $2\frac{3}{4}$, 107·3 halfpennies.

Unlike the halfpenny, however, the value of the quantum is not fixed, but is related to the frequency of radiation which, by its emission or absorption, causes the change in energy. This relationship is expressed as

$$E \quad = \quad h \quad \times \quad \nu$$

(Value of a quantum in joules)   (Planck's constant = 6·626 196 × 10⁻³⁴ J s)   (Frequency of radiation/s⁻¹ or Hz)

or, in terms of the wavelength (λ) of the radiation and the velocity of light (c), as

$$E = \frac{hc}{\lambda}$$

so that it is a simple matter to calculate the value of the quantum corresponding to any known frequency.

This idea was originally developed by Planck in considering a vibrating body changing in energy by emitting or absorbing radiation of a frequency equal to the frequency of the vibrating body, but Einstein showed that the idea held more generally and that if the energy of a body changed from a value E_1 to a value E_2 by emission or absorption of radiation of frequency ν, then

$$E_1 - E_2 = nh\nu = \frac{nhc}{\lambda}$$

where n is an integer. It was this generalised statement of the quantum theory that was used by Bohr in his interpretation of spectra.

5 Bohr's interpretation of spectral series   Rutherford assumed that the extra-nuclear electrons circulated round the nucleus of an atom in orbits, rather as planets circulate round the sun, and atoms were pictured as minute solar systems. The electrical forces of attraction between the negatively charged electrons and the positively charged nucleus were just counterbalanced by centrifugal forces.

Bohr pointed out, however, that electrons, i.e. charged particles, could not circulate in an orbit without having a corresponding acceleration towards the centre of the orbit, and, according to the accepted electrodynamic theory of the time, an electric charge must radiate energy when it is accelerated.

If Rutherford's idea is correct, then, an atom would radiate energy continuously. In doing so an atom would undergo spontaneous destruction. Moreover, the continuous emission of radiation would not account for the formation of line spectra.

To deal with these difficulties, Bohr put forward suggestions which, in effect, deny the truth of older electrodynamic theories as applied to the motion of electrons. These are:

(*a*) that the extra-nuclear electrons in an atom could only rotate in certain selected orbits and that they did not, in these orbits, radiate energy. Such orbits were called *stationary states*, and

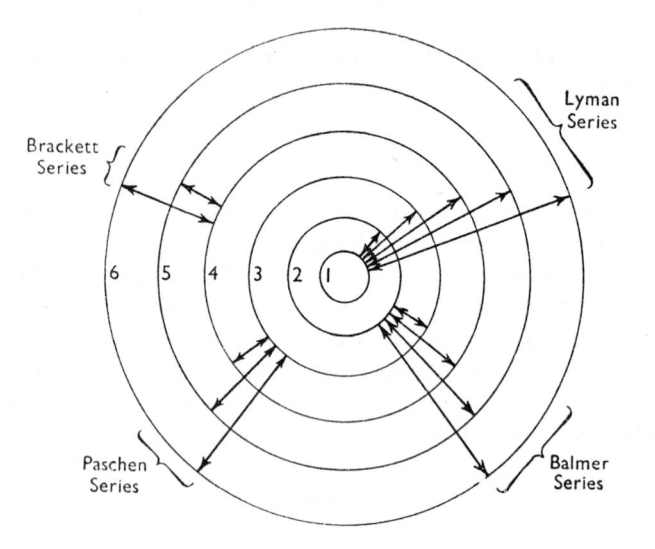

Fig. 3. Illustration of the energy changes which an electron can undergo in moving from one energy level to another in an atom. The diagram is not to scale, for the radii of the various stationary states shown are, in fact, proportional to the squares of the principal quantum numbers allotted to them

(*b*) that each stationary state corresponds to a certain energy level, i.e. that an electron in a certain stationary state had a certain energy, and that emission of radiant energy was caused by the movement of an electron from one stationary state to another of lower energy. Conversely, absorption of energy took place by an electron moving into a stationary state of higher energy.

Bohr now applied the ideas of the generalised quantum theory (page 11) to the change in energy when an electron moves from one stationary state to another. Thus, if the energy of one stationary state is given as E_1, and that of the next stationary state with lower energy as E_2, an electron passing from the first to the second would cause an energy change of $E_1 - E_2$, and an emission of radiation of frequency, ν, where

$$E_1 - E_2 = h\nu$$

Similarly, absorption of radiation of this frequency would cause an electron to pass from energy level E_2 to energy level E_1.

On this view the series observed in the line spectrum of hydrogen are explained by the various limited energy changes which an electron can undergo in moving between the various stationary states, which are characterised by principal quantum numbers of 1, 2, 3, ... The general idea is made clear by a study of Fig. 3 which shows the various energy changes leading to the various lines in the spectrum.

The atom is in its normal or ground state when the electron is in the stationary state of least energy; when in any other state the atom is said to be excited. On excitation the electron moves into stationary states of higher energy content, and it is the return of the electron to stationary states of lower energy which results in the emission of radiant energy and the formation of spectral lines. It is like lifting a ball up and letting it fall again, both the lifting and falling being done in definite stages.

For the hydrogen spectrum, the Bohr theory is able to account for the observed spectral lines and series in detail and with accuracy. The radii and energies of the various stationary states can easily be calculated as shown in the Appendix (page 228). Once the energies of the various stationary states are known, it is a simple matter to calculate the energy change when an electron passes from one energy state to another, and to relate the various energy changes to the observed spectral lines. Figure 4 shows how this is done for the hydrogen spectrum.

6 Units used Various units can be used to denote the energy changes involved. Energy change and wavelength (λ) or wave number (σ) are related (page 11) by the expression

$$E = \frac{hc}{\lambda} = hc\nu$$

so that the energy change corresponding to a particular wavelength or wave number is easily calculated. The value of Planck's constant,

h, is 6·626 2 × 10⁻³⁴ J s, and the value of the velocity of light, c, is 2·997 925 × 10⁸ m s⁻¹, so that

$$\frac{\text{Energy change}}{\text{(in joules)}} = \frac{1·986\ 495 \times 10^{-25}}{\text{wavelength (in m)}} = 1·986\ 495 \times 10^{-25} \times \text{wave-number (in m}^{-1})$$

A wavelength of 1 m, or a wave number of 1 m⁻¹, represents, therefore, an energy change of 1·986 495 × 10⁻²⁵ J.

This refers to the emission or absorption of energy by a single atom or molecule. On a mole basis, 6·022 169 × 10²³ atoms or molecules

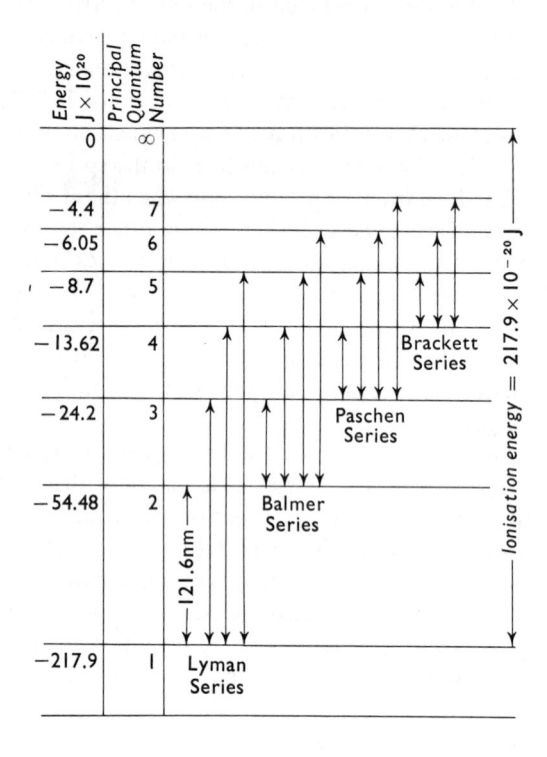

Fig. 4. Some of the energy levels in the hydrogen atom.

are involved so that the corresponding energy change becomes 1·196 291 × 10⁻¹ J mol⁻¹. This is equal to 2·8594 cal mol⁻¹.

Electron volts may also be used as units of energy instead of joules or calories. One electron volt is the energy gained by an electron in passing through a potential difference of 1 volt; it is equal to 1·602 10 × 10⁻¹⁹ J, 23·069 2 kcal mol⁻¹ or 96·487 kJ mol⁻¹.

In the hydrogen spectrum (Fig. 4) the first line in the Lyman series has an observed wavelength of 121·6 nm. This is brought about by an electron moving from a stationary state of energy $217·9 \times 10^{-20}$ J to one of $54·48 \times 10^{-20}$ J. The energy change of $163·42 \times 10^{-20}$ J leads to a spectral line of wavelength λ given by:

$$\lambda = \frac{1·986\ 495 \times 10^{-25}}{163·42 \times 10^{-20}} = 1·216 \times 10^{-7}\ \text{m} = 121·6\ \text{nm}.$$

QUANTUM NUMBERS

The extension of Bohr's ideas came about as a result of more detailed investigation of spectra using equipment with higher resolving power. Many single lines in spectra were found to consist of a number of very closely related lines. Moreover, many single lines were found to split when the source of radiation was placed in a magnetic field (the *Zeeman effect*) or in an electrical field (the *Stark effect*).

The resulting fine structure of spectra necessitates an increase in the number of possible orbits in which an electron can be said to exist within an atom. In other words, it is necessary to allow for more possible energy changes within an atom to account for the greater number of observed spectral lines.

The term quantum number is used to label the various energy levels, and four types of quantum number are necessary.

7 Principal quantum number The number allotted to Bohr's original stationary states, visualised as circular orbits, is called the principal quantum number. The innermost orbit, i.e. that nearest the nucleus, has a principal quantum number of 1: the second orbit has a quantum number of 2, and so on. Alternatively letters are used to characterise the orbits, the first being referred to as the K orbit, the second as the L, the third as the M, and so on. The choice of letters originates from Moseley's work on the X-ray spectra of the elements. He called groups of lines in the spectra the $K, L, M, N, \ldots$ groups (page 23).

The number of electrons in an atom which can have the same principal quantum number is limited and is given by $2n^2$ where n is the principal quantum number concerned. Thus

Principal quantum number (n)	1	2	3	4
Letter designation	K	L	M	N
Maximum number of electrons	2	8	18	32

8 Subsidiary quantum number For each value of the principal quantum number there are several closely associated orbits, so that the principal quantum number represents a *group* or *shell* of orbits. The subsidiary quantum number symbolised by *l*, is used to label the subsidiary orbits within a shell. They may be visualised, in a simple way, as elliptical, rather than circular, orbits. For this reason, the subsidiary quantum number is sometimes referred to as the *azimuthal* quantum number.

In any one shell, having the same principal quantum number, the various subsidiary orbits are denoted as the 1, 2, 3, 4, ... or the *s*, *p*, *d*, *f*, ... orbits. The letters originate from the sharp, principal, diffuse and fundamental series of lines in spectra (page 9).

The number of subsidiary orbits in any one shell is limited and in dealing with the structures of atoms in their ground or normal states it is only necessary to consider the following subsidiary orbits:

1*s*	2*s*	3*s*	4*s*	5*s*	6*s*	7*s*
	2*p*	3*p*	4*p*	5*p*	6*p*	
		3*d*	4*d*	5*d*	6*d*	
			4*f*	5*f*		

The 1*s* orbit may be referred to as the *K*1 orbit but the commonest usage employs numbers for the principal quantum number and letters for the subsidiary quantum number.

It is the difference in energy level between the orbits in any one shell that immediately allows more energy changes to take place within an atom. Some of the possible energy changes in a sodium atom are shown, for example, in Fig. 5, and each change gives a corresponding line in the spectrum. The figure also illustrates the origin of the sharp, principal, diffuse and fundamental series.

Some of the theoretically possible energy changes in an atom do not take place and there are no corresponding lines in the spectrum. Spectroscopists apply empirical *selection rules* in determining which energy changes are possible and which are forbidden.

9 Spin quantum number The *Pauli* or *Exclusion* principle states that all the electrons in any atom must be distinguishable, or that no two electrons in a single atom can have all their quantum numbers alike. This is a most important principle but it cannot be derived from any fundamental conceptions. Its application involves the conception of electron spin, in which an electron is assumed to be capable of spinning on its axis.

Electron spin was first postulated in 1925 by Uhlenbeck and Goudsmit to account for the splitting of many single spectral lines into double lines when examined under a spectroscope of high resolving power. In effect, the conception of electron spin makes a further quantum number necessary for the spin can be in one of two directions. The spin quantum number can, therefore, have one of two values, generally written as $s=+\frac{1}{2}$ or $s=-\frac{1}{2}$.

The conception of electron spin is important in applying the Pauli principle. The shell of principal quantum number 1 can hold two electrons. As these cannot be exactly alike, because of the Pauli principle, it is assumed that they have different or opposed spins. In general, if two electrons occupy the same orbit they must have

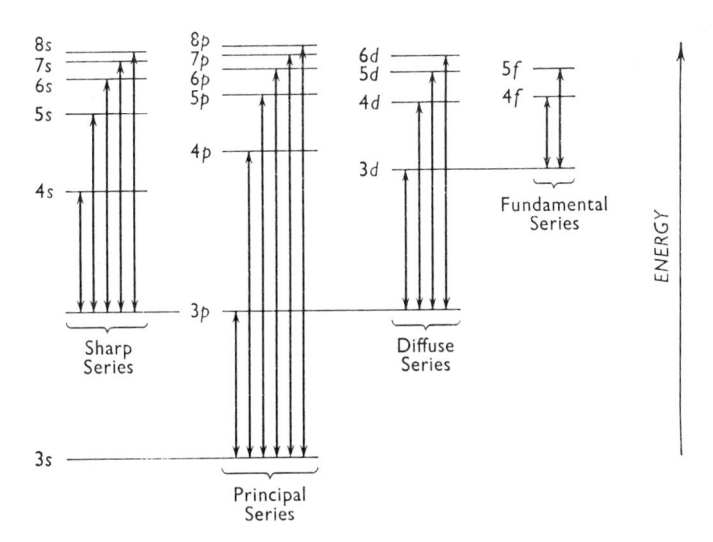

Fig. 5. Some of the possible energy changes within a sodium atom giving rise to the sharp, principal, diffuse and fundamental series of lines in the spectrum

different spins (they are said to be 'paired') and, as electrons in the same orbit can only differ by having different spins, it follows that no orbit can contain more than two electrons.

10 Magnetic quantum number For electrons in orbits of principal quantum number greater than 1, a further complication arises. The shell with principal quantum number of 2 can contain a maximum of eight electrons. Of these, two, with opposed spins, will be in the 2s level. The remaining six will be in the 2p level, but for all six to be different it is necessary to subdivide the 2p level into three so that each of the three subdivisions may contain two electrons. These three 2p orbits may be envisaged as being in different planes and can be denoted as $2p_x$, $2p_y$ and $2p_z$ orbits (Fig. 35). Each of the three orbits can contain two electrons with opposed spins.

The subdivision of p orbits into three, and a similar subdivision of d orbits into five (see page 78) and of f orbits into seven, necessitates a fourth quantum number known as the magnetic quantum number and symbolised by m.

For simple purposes the subdivision of p orbits is not of great importance because the energy of an electron in any one of the three p orbits is the same unless the atom is placed in a strong magnetic field. The p orbits are said to be *degenerate* or to have a degeneracy of three. In a magnetic field, the three orbits take up different positions with respect to the lines of force of the field and attain slightly different energy levels. This accounts for the splitting of spectral lines when the source of emission is placed in a magnetic field (the Zeeman effect) or in an electric field (the Stark effect).

11 Summary To sum up, four quantum numbers are required to characterise completely any particular electron in a particular orbit. In a simple way this corresponds to a normal post office address. To characterise a particular Mr X it is necessary to allot a particular address to him, e.g. Mr X, 114, High Street, Eton, Bucks. The county corresponds to the principal quantum number, the town to the subsidiary quantum number, the street to the spin quantum number, and the street number to the magnetic quantum number.

The idea of quantum numbers has been presented from a pictorial point of view. What quantum numbers are allowed arises, mathematically, from the solution of the Schrödinger equation (page 69) and those quantum numbers which are allowed can be summarised as follows:

(*a*) n = principal quantum number. The allowed values are 1, 2, 3, 4, . . .

(*b*) l = subsidiary quantum number. The allowed values of l depend on the value of n. When n is 1, l is 0, i.e. there are only s electrons in the shell with principal quantum number 1.

When n is 2, l can be 0 or 1, i.e. the shell with principal quantum number of 2 contains both s and p electrons.

When n is 3, l can be 0, 1 or 2, i.e. s, p and d electrons. When n is 4, l may be 0, 1, 2 or 3, i.e. s, p, d and f electrons.

It will be seen that $l = 0$ for an s electron, 1 for a p electron, 2 for a d electron and 3 for an f electron.

(c) m = magnetic quantum number. For an s electron with $l=0$, m is 0. For a p electron, with $l=1$, m can be $-1, 0$ or $+1$. For a d electron with $l=2$, m can be $-2, -1, 0, +1$ and $+2$, and so on. In general, m can have $(2l+1)$ different values.

(d) s = spin quantum number. The allowed values are $+\frac{1}{2}$ and $-\frac{1}{2}$.

The maximum number of electrons with the same principal quantum number can be summarised as follows, bearing in mind the fact that no two electrons in the same atom can have the same values for the four quantum numbers.

n	1	2				3								
l	0	0	1			0	1			2				
m	0	0	-1	0	$+1$	0	-1	0	$+1$	-2	-1	0	$+1$	$+2$
s	$+\frac{1}{2}-\frac{1}{2}$	$+\frac{1}{2}-\frac{1}{2}$	$+\frac{1}{2}-\frac{1}{2}$	$+\frac{1}{2}-\frac{1}{2}$	$+\frac{1}{2}-\frac{1}{2}$	$+\frac{1}{2}-\frac{1}{2}$	$+\frac{1}{2}-\frac{1}{2}$	$+\frac{1}{2}-\frac{1}{2}$	$+\frac{1}{2}-\frac{1}{2}$	$+\frac{1}{2}-\frac{1}{2}$	$+\frac{1}{2}-\frac{1}{2}$	$+\frac{1}{2}-\frac{1}{2}$	$+\frac{1}{2}-\frac{1}{2}$	$+\frac{1}{2}-\frac{1}{2}$

$$\underbrace{}_{2} \quad \underbrace{}_{8} \quad \underbrace{}_{18}$$

3 The Periodic Table

1 Historical introduction Once the relative atomic masses of some elements became known, interesting numerical relationships were soon noticed, and these eventually led to the periodic table in which the elements are arranged in order of increasing relative atomic mass.

Even before many relative atomic masses were known with great precision, Dobereiner noticed, in 1829, that certain groups of three chemically similar elements had values which were approximately in arithmetic progression. Such groups became known as Dobereiner's triads and are exemplified, using modern values, by

Chlorine	Bromine	Iodine	Lithium	Sodium	Potassium
35·5	79·9	126·9	6·9	23	39·1
Sulphur	Selenium	Tellurium	Calcium	Strontium	Barium
32	79	127·6	40·1	87·6	137·4

Other similar, but, at the time, mysterious numerical relationships using both relative atomic and equivalent masses also came to light, and led to Newlands' Law of Octaves in 1864. Newlands arranged all the elements he knew in ascending order of relative atomic mass and assigned to the elements a series of ordinal numbers which he called atomic numbers. He then noticed that elements with similar chemical properties had atomic numbers which differed by seven or some multiple of seven. In other words, Newlands discovered that the chemical properties of elements were often found to be similar for every eighth or sixteenth elements, like the notes in octaves of music.

Newlands' ideas were not, however, widely accepted and were subjected to some ridicule, until, in 1869, they were essentially restated by Mendeleef and by Lothar Meyer. The periodic recurrence of similar chemical properties when the elements are arranged in ascending order of relative atomic mass, which was the essential point of what Mendeleef called the law of periodicity, was supported by the periodic relationships noticed by Lothar Meyer in plotting the atomic volumes (relative atomic masses/densities) of the elements against their relative atomic masses (Fig. 6).

2 Mendeleef's form of periodic table Mendeleef's original periodic table arrangements were, naturally, incomplete because he only knew a limited number of elements. He had, of necessity, to leave many gaps in his arrangement for elements not, at that time, discovered.

Mendeleef predicted both the eventual discovery of such elements and their probable properties, and it was a real triumph that many gaps he had originally left were, in his lifetime, filled by newly discovered elements having the properties he had predicted.

A modernised version of Mendeleef's table is given on the front end-paper and the following points are worthy of attention, for many of the original terms are still used in a modified setting.

(*a*) *Groups and periods.* The elements fall into nine vertical groups, and elements in the same group have certain chemical and physical similarities. In some groups the similarities are very marked; in others, less so.

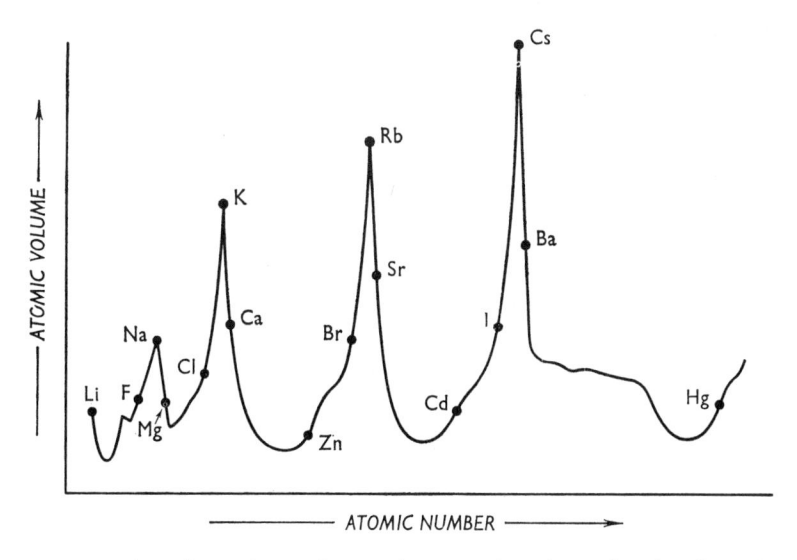

Fig. 6. Plot of atomic number against atomic volume for the elements showing some periodic relationships

The horizontal groups are known as periods, those containing eight elements being called short periods, and the others long periods.

(*b*) *Typical elements.* The elements in the short periods at the head of each vertical group are known as typical elements: their chemistry is, on the whole, typical of the group. Below the typical elements, the groups are usually subdivided into sub-groups, A and B. The similarity between elements in different sub-groups, but in the same group, is not very marked although it is noticeable that the sub-group A elements resemble the typical elements in the left-hand groups, whereas sub-group B elements are more like the typical elements in the right-hand groups.

(c) *Transitional elements*. The three groups of elements placed in Group 8 were originally called transition, or transitional, elements. They show some resemblance to the elements preceding them, and to those following them, and in that way they link up the first and second halves of the long periods. The term transition element is, however, now used in a wider sense, as explained on page 31.

(d) *Group valencies*. The numerical valency of an element is often equal either to the group number in which the element occurs, or to eight minus the group number. Such a relationship was expressed, in 1904, in Abegg's rule of eight. It is by no means universally true, but the number 8 plays an important part in valency considerations, as will be seen.

(e) *Anomalous placings*. In order to maintain the chemical similarities in the Mendeleef form of periodic table, a few elements have to be placed in their wrong relative atomic mass order. In the following pairs of elements, for example, the first element, with the lower relative atomic mass, is placed after the second in the periodic table (see p. 24),

Potassium	Argon	Nickel	Cobalt
39·10	39·95	58·71	58·93
Iodine	Tellurium	Protoactinium	Thorium
126·90	127·60	231	232·04

(f) *Position of hydrogen*. Hydrogen has some likenesses both to the alkali metals and the halogens and it can be placed at the head of Group 1 or Group 7, though neither place is entirely satisfactory for it on purely chemical grounds.

It can enter into chemical combination in a number of different ways (some of them unique) as described in Chapter 13.

(g) *Rare earths, lanthanons or lanthanides: actinons or actinides*. The rare earths, lanthanons or lanthanides form a group of 14 elements with very marked chemical similarity. All 14 elements occupy one single position in Group 3 in the periodic table. The actinons or actinides provide a similar series.

(h) *Noble gases*. The discovery of a whole new series of elements could not have been predicted from the early forms of the periodic table. The noble gases, discovered at the turn of the nineteenth and twentieth centuries, provided such a series which had to be fitted into the periodic table. This was conveniently done by placing them in

Group 0, a position which suggested a valency of zero in keeping with their chemical inactivity (page 36).

3 Thomsen-Bohr periodic table Very many variations on the periodic table theme have been put forward, each aiming to emphasise some particular feature. Perhaps the main disadvantages of the Mendeleef table are the division of groups into subgroups and the isolation of a limited number of transition elements in Group 8.

To overcome such difficulties Thomsen proposed a form of the periodic table, in 1885, which has been developed, by Bohr and others, into the arrangement shown on the front end-paper.

This form of the periodic table is more in keeping with the known atomic structures of the elements (page 28) and is the most commonly used periodic table arrangement. It can, of course, be modified in a number of ways so far as order of presentation is concerned.

4 X-ray spectra At first, there was no genuine understanding as to why an arrangement of elements in relative atomic mass order should fit elements into groups with similar chemical properties. The deeper significance of the arrangement came, first, from a study of X-ray spectra, and, more completely, from a knowledge of the arrangement of the electrons in different atoms (page 24).

When a solid anode is bombarded by cathode rays, X-rays are produced, and von Laue suggested, in 1912, that it might be possible to obtain diffraction patterns from X-rays if, in fact, they were wave-like in nature. Friedrich tested this suggestion experimentally, and found that a copper sulphate crystal could be used to diffract X-rays. This established the nature of X-rays as light-like radiation of very short wavelength, and also introduced the use of crystals, with a regular internal array of atoms, as diffraction gratings for radiation of short wavelength. It was, in fact, the beginning of X-ray crystallography.

The X-rays given out by any solid anode are of varying wavelength. Bombardment of every element gives a general, or white, X-radiation which shows up in an X-ray spectrum as a continuous background. Superimposed on this background, however, are a number of lines characteristic of the particular element being used as the anode. These lines occur in groups known as the $K, L, M, N, \ldots$ groups, and there are various lines within each group. The letters $K, L, M, N, \ldots$ were originally chosen so that there was room for expansion in either direction if new groups were discovered.

By measuring the wavelengths of corresponding lines in the X-ray spectra of as many elements as possible, Moseley discovered, in 1913, that the square root of the frequency of corresponding lines for different elements gave almost a straight line when plotted against the atomic numbers of the elements concerned (Fig. 7).

This suggested, very strongly, that the atomic number of an element is really of more significance than the relative atomic mass in the periodic table arrangement, and, on this basis, Moseley was able to make some adjustments in the older periodic table orders. Cobalt, for example, has a higher relative atomic mass than nickel (page 22) and

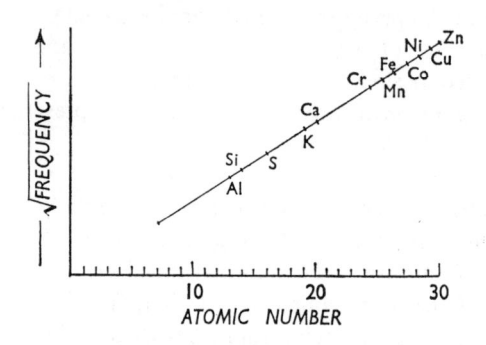

Fig. 7. The relationship between the atomic numbers of some simple elements and the square root of the Kα lines in their X-ray spectra

ought to follow it in the periodic table so far as relative atomic masses are concerned. Moseley found, however, that X-ray spectral lines of nickel had higher frequencies than those of cobalt, and he therefore gave nickel the higher atomic number.

5 The arrangement of electrons in orbits   The arrangement of electrons in an atom in its normal or ground state is that which makes its energy a minimum, i.e. it is the most stable arrangement, within the limitations of allowable quantum numbers mentioned on page 18.

The energy levels of the various orbits which an electron might occupy can be obtained from spectroscopic data, and the detailed arrangements of electrons in atoms can be built up from a knowledge of these energy levels, from the position of the atom concerned in the periodic table, and from the allowable quantum numbers.

The relative energy levels of orbits in an atom, as obtained from spectroscopic measurements, are shown in Fig. 8. The precise positioning of one orbit to another depends on the atomic number of the element concerned. The order given is that for elements of low atomic number, and it changes to some extent with elements of higher atomic number. This is because of the changing nuclear attraction for the electrons as the positive charge on the nucleus changes.

In Fig. 8, each circle represents an orbit which can be occupied either by a single electron or by two electrons with different spins. The circles enclosed within a rectangle represent orbits of equal energy under normal circumstances.

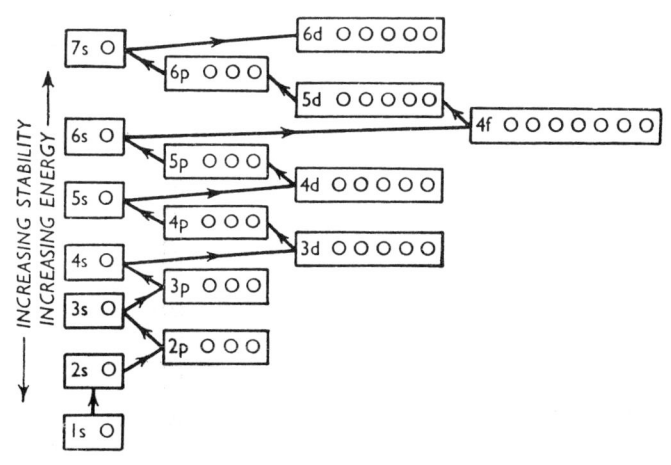

Fig. 8. The approximate relative energy levels of orbits in an atom

In passing along the periodic table of the elements the electrons in the atoms concerned occupy the orbits in energy order, as shown by the arrows in Fig. 8. The correct energy order might, perhaps, be remembered more easily by thinking in terms of Fig. 9. Alternatively, the correct sequence can be obtained by considering the sum of the principal and subsidiary quantum numbers, $(n+l)$, for the various possible orbits. The orbit with the lowest $(n+l)$ value fills first, i.e. 1s $(n+l=1)$ before 2s $(n+l=2)$. When two orbits have the same $(n+l)$ value the one with the lower n value fills first, i.e. 3p $(n+l=4)$ before 4s $(n+l=4)$.

The one electron in a hydrogen atom will occupy the most stable orbit, i.e. the 1s orbit, and the second electron in the helium atom will occupy the same orbit but will have a different spin. The 1s orbit is now full. A third electron will occupy the next most stable orbit, i.e. the

2s orbit, and so on. The best 'seats' are occupied first. This building-up process is often referred to as the *aufbau* principle.

The application of the principle is not absolutely regular, as will be seen from a careful study of the arrangements of electrons in the atoms of the elements in their ground or normal states summarised on the back end-paper. These electronic arrangements are derived from spectroscopic measurements. They do not differentiate between the three *p*, five *d* or seven *f* levels, and they are commonly written in the form $1s^2$, $2s^2$, $2p^6$, $3s^2$, $3s^6$, which is the argon arrangement.

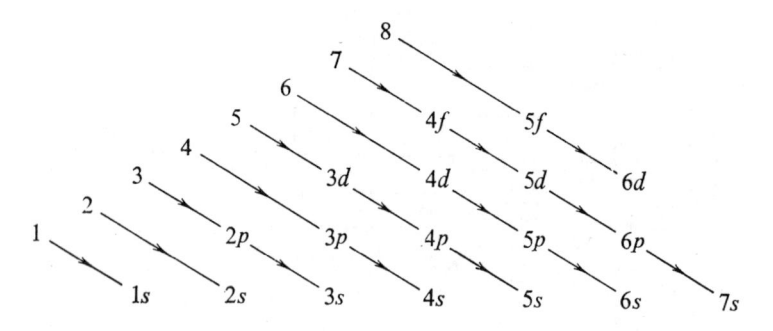

Fig. 9. General order in which orbits in an atom are occupied. Details of irregularities are given in Section 6.

6 Deviations from the Aufbau principle The actual arrangement of electrons in an atom as indicated by spectroscopic studies is, for a few elements, slightly different from the arrangement of electrons predicted by the application of the aufbau principle. These slight deviations from regularity involve the placing of one or two *ns*-electrons in an $(n-1)\,d$ level. The energy difference between such *s* and *d* levels is small, so that there is very little to choose, from an energy point of view, between one structure and the other.

The more important deviations from regularity are summarised as follows, only the relevant outer electrons being given:

	Expected arrangement	Actual arrangement
24 Cr	$3d^4$ $4s^2$	$3d^5$ $4s$
29 Cu	$3d^9$ $4s^2$	$3d^{10}$ $4s$
42 Mo	$4d^4$ $5s^2$	$4d^5$ $5s$
46 Pd	$4d^8$ $5s^2$	$4d^{10}$
47 Ag	$4d^9$ $5s^2$	$4d^{10}$ $5s$
78 Pt	$5d^8$ $6s^2$	$5d^9$ $6s$
79 Au	$5d^9$ $6s^2$	$5d^{10}$ $6s$

It will be seen that the deviations tend to occur when a *d* level is either almost full or half-full.

7 The rule of maximum multiplicity A still more detailed arrangement of electrons can be predicted by application of the rule of maximum multiplicity. This empirical rule, suggested by Hund, states that the distribution of electrons in a free atom between the three *p*, five *d* and seven *f* orbits is such that as many orbits as possible are occupied by single electrons before any pairing of electrons with opposed spins takes place. Thus if three electrons are to occupy the three *p*-orbits in any one shell one will go into each of the three available orbits. This can be interpreted as meaning that electrons

	1s	2s	$2p_x$	$2p_y$	$2p_z$
1 H	↓				
2 He	↓↑				
3 Li	↓↑	↓			
4 Be	↓↑	↓↑			
5 B	↓↑	↓↑	↓		
6 C	↓↑	↓↑	↓	↓	
7 N	↓↑	↓↑	↓	↓	↓
8 O	↓↑	↓↑	↓↑	↓	↓
9 F	↓↑	↓↑	↓↑	↓↑	↓
10 Ne	↓↑	↓↑	↓↑	↓↑	↓↑

Fig. 10. The distribution of electrons between three 2p orbits in the elements from H—Ne

repel each other and keep as far away from each other as is possible.

The distribution of electrons between three 2*p* orbits and five 3*d* orbits is shown in Figs. 10 and 11, where the detailed electronic structures of elements of atomic numbers 1–10 and 19–30 are shown. No absolute significance is to be attached to an arrow pointing up or down. It is the relative spins of electrons, and whether they are paired or not, that matters.

These more detailed arrangements are important, but are not always necessary. There are, in general, three ways of expressing an electronic arrangement within an atom. Fluorine, for example, has nine electrons. The arrangement of the electrons can be written as 2.7, showing, simply, the number of electrons having different principal quantum numbers. In more detail, the arrangement can be written as $1s^2$, $2s^2$, $2p^5$, and the arrangement given in Fig. 10, showing the spin and the sub-division of p orbits, is still more detailed.

The method most appropriate to the problem being considered should be chosen; it is wasteful to take a sledge-hammer to crack a walnut.

	1s	2s	2p	3s	3p	←		3d		→	4s
19 K	2	2	6	2	6						↓
20 Ca	2	2	6	2	6						↓↑
21 Sc	2	2	6	2	6	↓					↓↑
22 Ti	2	2	6	2	6	↓	↓				↓↑
23 V	2	2	6	2	6	↓	↓	↓			↓↑
24 Cr	2	2	6	2	6	↓	↓	↓	↓	↓	↓
25 Mn	2	2	6	2	6	↓	↓	↓	↓	↓	↓↑
26 Fe	2	2	6	2	6	↓↑	↓	↓	↓	↓	↓↑
27 Co	2	2	6	2	6	↓↑	↓↑	↓	↓	↓	↓↑
28 Ni	2	2	6	2	6	↓↑	↓↑	↓↑	↓	↓	↓↑
29 Cu	2	2	6	2	6	↓↑	↓↑	↓↑	↓↑	↓↑	↓
30 Zn	2	2	6	2	6	↓↑	↓↑	↓↑	↓↑	↓↑	↓↑

Fig. 11. The distribution of electrons between five 3d orbits in the elements from K—Zn

8 Types of element An examination of the electron structures of the atoms reveals five types of element, and this sub-division is very useful. The five types are summarised below; and shown in the periodic table arrangement given in Fig. 12 on page 29.

(a) The noble gases. The arrangement of the electrons in the noble gases is as follows:

		1	2		3			4				5			6	
		s	s	p	s	p	d	s	p	d ·	f	s	p	d	s	d
2	He	2														
10	Ne	2	2	6												
18	A	2	2	6	2	6										
36	Kr	2	2	6	2	6	10	2	6							
54	Xe	2	2	6	2	6	10	2	6	10		2	6			
86	Rn	2	2	6	2	6	10	2	6	10	14	2	6	10	2	6

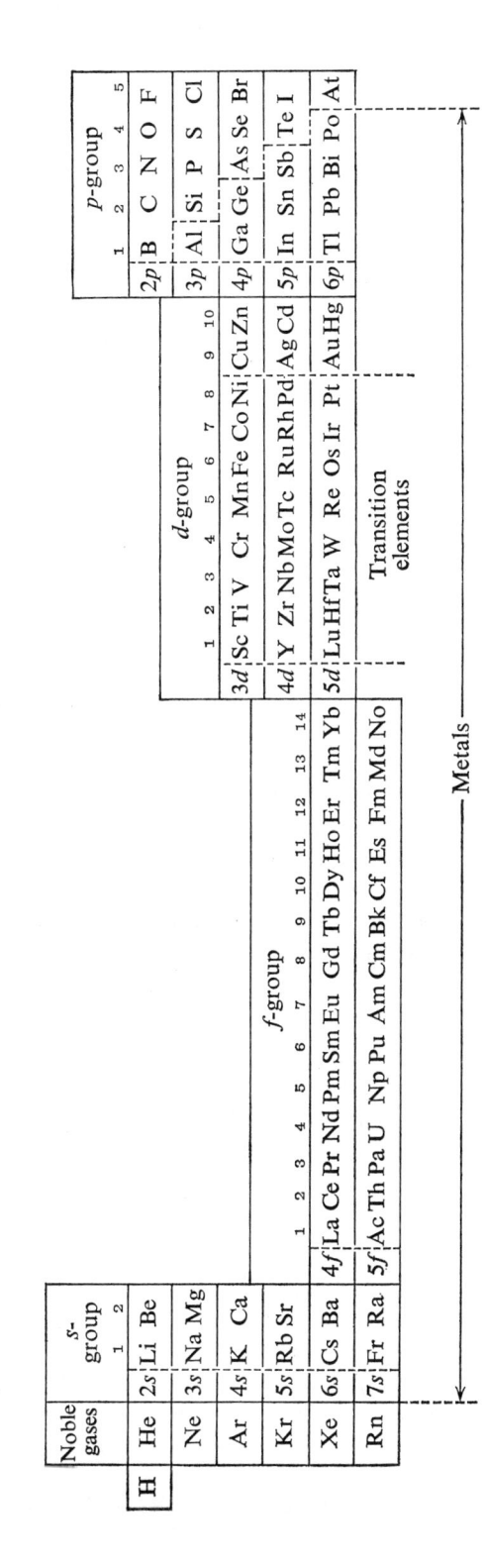

Fig. 12. *Modern arrangement of periodic table showing s–, p–, d– and f– groups of elements*

Every orbit which is occupied at all is fully occupied and it is this unique arrangement of electrons which accounts for the chemical inactivity of the noble gases (see, however, page 36).

(b) *s-block elements*. These elements, found in Groups 1A and 2A, have atoms containing either one or two electrons in the outermost *s* orbit, all other orbits which are occupied being fully occupied. The structures of the atoms concerned may be summarised as follows:

Li	2.1	Be	2.2
Na	2.8.1	Mg	2.8.2
K	2.8.8.1	Ca	2.8.8.2
Rb	2.8.18.8.1	Sr	2.8.18.8.2
Cs	2.8.18.18.8.1	Ba	2.8.18.18.8.2
Fr	2.8.18.32.18.8.1	Ra	2.8.18.32.18.8.2

The chemical similarity within the group is due to the fact that all the elements in the same group have the same number of electrons in the outermost orbit, this number being equal to the group number. Atoms with the same number of electrons in their outermost orbits would be expected to be chemically similar, for when two atoms approach each other prior to chemical combination it will be the outermost electrons of the atoms which interact to the greatest extent and participate in chemical bonding.

For this reason, the outermost electrons are sometimes referred to as *valency electrons*, and, for many purposes, the other electrons can be disregarded. The structures of the sodium and calcium atoms can usefully be written, for example, as follows:

$$Na^x \qquad Ca^x_x$$

The *s*-block elements, excluding hydrogen, are strongly electropositive metals, which form colourless ions and which exert a fixed valency, equal to their group number, in almost all their compounds. Hydrogen can be regarded as an *s*-block element, but it is generally best to exclude it because its chemical functioning is unique (page 141).

(c) *p-block elements*. These elements occur in Groups 3–7B. They have atoms in which the outermost *p* orbit is filling up from one to five electrons, the total number of electrons in the outermost shell rising from three for Group 3 elements to seven for Group 7 elements.

Of the 25 elements concerned, 15 would generally be regarded as non-metallic and ten as metallic, though the diagonal dividing line must not be taken absolutely rigidly because the distinction between a metal and a non-metal is not absolutely clear cut.

```
Group no.   3    4    5    6    7
            B    C    N    O    F
            Al | Si   P    S    Cl
            Ga   Ge | As   Se   Br        Non-metals
Metals      In   Sn   Sb | Te   I
            Tl   Pb   Bi   Po | At
```

The p-block elements generally exert valencies equal to the group number or to the group number minus two.

(d) *d-block elements.* In these elements an inner $3d$, $4d$ or $5d$ orbit fills up from one to ten electrons so that there are three series of ten elements — Sc to Zn, Y to Cd and La to Hg (omitting the rare earths). In the first eight elements in each of these series, the $3d$, $4d$ or $5d$ orbit contains one to eight electrons, i.e. it is occupied but not fully. These elements are the transition elements. They form coloured ions, which are paramagnetic (page 64); exhibit variable valencies, which may differ from each other by one unit; form many complexes; and often possess marked catalytic activity.

Cu and Zn, Ag and Cd, and Au and Hg contain 10 electrons in the $3d$, $4d$ or $5d$ orbit, i.e. the d orbit is full. To that extent they are different from the transition elements. Cu, Ag and Au (the coinage metals) do, however, have many chemical characteristics in common with the transition elements and are often referred to as transition elements. Zn, Cd and Hg may also be so called, even though their chemical characteristics are rather different. There is, in fact, some difference of opinion as to the precise meaning to be given to the term transition element.

(e) *f-block elements.* There are two series of f-block elements in which the $4f$ and $5f$ orbits fill up from 1 to 14. The first series contains the rare-earths or lanthanons or lanthanides. It is because there are the same number of electrons in the two outermost shells in all these 14 elements that they are so very much alike.

The second series contains the actinons or actinides, and, again, there is remarkable chemical similarity.

4 Types of Chemical Bond

1 Introduction The peculiar stability of the electronic structures of the inert gases, as shown by their lack of chemical reactivity, led to the first modern hypothesis regarding the mechanism of formation of chemical bonds. The underlying idea was that an atom combined with another atom in such a way that both atoms achieved a noble gas electronic structure. In this way, compound formation was represented as a process resulting in greater stability.

This is a simple idea and the first suggestions as to the mechanism of chemical bonding were simple, too. As will be seen, however, they have required constant extension. The simple ideas will be introduced here; some of the extensions are discussed in later chapters.

Different types of chemical bond were originally thought of as completely distinct, and, in the early stages, they are best dealt with in this way. It is important to realise, however, that these distinctions are now made as a matter of convenience and that most actual chemical bonds are intermediate in type.

2 The ionic bond Kossel suggested, in 1916, that those elements placed just before a noble gas in the periodic table could attain a noble gas structure by gaining electrons and forming negatively charged ions. Thus chlorine, with a structure 2.8.7, could become a chloride ion, Cl^-, with a structure 2.8.8 by gaining an electron.

Similarly, an element placed just after a noble gas could achieve noble gas structure by losing electrons and forming positively charged ions, e.g. sodium, 2.8.1, could form a sodium ion, Na^+, with the inert gas structure, 2.8.

Both the chloride and sodium ions would have stable structures, and by combining would form sodium chloride. For two ions, sometimes known as an *ion-pair*, the state of affairs could be represented, showing only the outermost, or valency, electrons as:

$$Na\cdot + {}^{\times}_{\times}\!\overset{\times\times}{\underset{\times\times}{Cl}}{}^{\times}_{\times} \rightarrow [Na]^+ \left[{}^{\times}_{\times}\!\overset{\times\times}{\underset{\times\times}{\overset{\bullet}{Cl}}}{}^{\times}_{\times}\right]^-$$

the two ions being held together by electrostatic attraction. When, as in practice, many ions are involved, they are held together, by electrostatic forces, within a crystal known as an *ionic crystal* (page 55).

Many simple compounds can be formulated as ionic compounds

containing ionic bonds. Calcium bromide and potassium sulphide provide typical examples:

$$[\text{Ca}]^{2+} \quad \begin{matrix} \left[\overset{\times\times}{\underset{\times\times}{\overset{\bullet}{_\times}\text{Br}\overset{\times}{_{}}}} \right]^{-} \\ \left[\overset{\times\times}{\underset{\times\times}{\overset{\bullet}{_\times}\text{Br}\overset{\times}{_{}}}} \right]^{-} \end{matrix} \qquad \begin{matrix} [\text{K}]^{+} \\ [\text{K}]^{+} \end{matrix} \quad \left[\overset{\times\times}{\underset{\times\times}{\overset{\bullet}{_{}}\text{S}\overset{\times}{_{}}}} \right]^{2-}$$

Calcium bromide, CaBr₂   Potassium sulphide, K₂S

The commonest and most stable ionic compounds are formed between elements preceding, and elements following, a noble gas in the periodic table. Elements preceding a noble gas form negatively charged ions (anions) by gaining electrons; such elements are said to be *electronegative*. Elements following a noble gas, and readily losing electrons to form positively charged ions, are said to be *electropositive*.

Ionic compounds are invariably electrolytes, and are, generally, hard solids with high melting and boiling points, insoluble in benzene and other organic solvents but soluble in water (page 51).

3 The covalent bond There are many substances which cannot be formulated with ionic bonds, either because they are non-electrolytes, e.g. tetrachloromethane, or because the atoms bonded together are the same, so that neither would be expected to transfer an electron to the other, e.g. chlorine, Cl_2.

To account for the formation of such molecules, Lewis, in 1916, suggested that atoms might attain a noble gas structure, not by complete transference of electrons, as in ionic bonding, but by sharing electrons.

On this idea the chlorine and hydrogen molecules are represented as:

$$\overset{\bullet\bullet}{\underset{\bullet\bullet}{\overset{\bullet}{_\bullet}\text{Cl}\overset{\bullet}{_\bullet}}} \overset{\times\times}{\underset{\times\times}{\text{Cl}\overset{\times}{_\times}}} \quad \text{and} \quad \text{H}\overset{\bullet}{_\times}\text{H}$$

one electron from each atom being held in common by both. The shared pair of electrons constitutes what is known as a covalent bond.

In tetrachloromethane and methane the molecules are represented as:

$$\begin{matrix} & \overset{\bullet\bullet}{\bullet}\text{Cl}\overset{\bullet\bullet}{\bullet} & \\ \overset{\bullet\bullet}{\bullet}\text{Cl}\overset{\bullet}{_\times} & \text{C} & \overset{\times}{_\bullet}\text{Cl}\overset{\bullet\bullet}{\bullet} \\ & \overset{\bullet\bullet}{\bullet}\text{Cl}\overset{\bullet\bullet}{\bullet} & \end{matrix} \qquad \begin{matrix} \text{H} \\ \text{H}\overset{\times}{_\times}\text{C}\overset{\times}{_\bullet}\text{H} \\ \text{H} \end{matrix}$$

and the double and triple bonds of organic compounds, as exemplified by ethene and ethyne, are written as:

$$\begin{array}{cc} H & H \\ C::C \\ H & H \end{array} \qquad H:C:::C:H$$

Ethene   Ethyne

A shared pair of electrons represents one covalent bond, and it is convenient to denote such a single bond by a single line. Molecules of ammonia, hydrogen chloride, oxygen and nitrogen can, therefore, be written either in terms of shared electrons or in terms of single, double or triple bonds, as shown:

$$H:\underset{H}{\overset{\times\times}{N}}:H \quad \text{or} \quad H-\underset{H}{N}-H \qquad H:\overset{\times\times}{\underset{\times\times}{Cl}}: \quad \text{or} \quad H-Cl$$

Ammonia   Hydrogen chloride

$$:O::O: \quad \text{or} \quad O=O \qquad :N:::N: \quad \text{or} \quad N\equiv N$$

Oxygen   Nitrogen

A covalent bond is directed in space so that the atoms in a covalent compound are linked in a definite position in relation to each other, and the molecules formed may exist as distinct particles. The atoms in a water molecule, for example, are arranged as shown:

$$\begin{array}{c} H \\ O \\ H \end{array} \quad \text{or} \quad O\!\!\begin{array}{c} H \\ \\ H \end{array}$$

with a bond angle of 104°31′. Similarly, when carbon forms four single covalent bonds the bonds are arranged tetrahedrally. The molecule of methane, therefore, although commonly represented, for convenience, as flat, is really three-dimensional with the four hydrogen atoms at the corners, and the carbon atom at the centre, of a tetrahedron.

The fact that many molecules are three-dimensional must not be forgotten. They are often portrayed as flat simply because it is so much more difficult to show their spatial arrangement on paper, and because, for some purposes, this arrangement is not important.

4 The dative bond The shared pair of electrons of a covalent bond may also be formed by one of the two bonded atoms providing both electrons. In such a case the bond is sometimes called a dative bond, but as it is just like a covalent bond, once it is formed, the two are not always distinguished in bond diagrams.

The atom providing the two electrons to make up the dative bond is known as the *donor*. It must, of course, have an 'unused' pair of electrons available, and such a pair is referred to as a *lone pair*. The atom sharing the pair of electrons from the donor is known as the *acceptor*.

When it is not necessary to distinguish between a dative bond and a covalent bond the — symbol is used for both. Two other symbolisms to represent a dative bond are, however, in use and have certain points in their favour.

The first shows a dative bond between atoms A and B as $A \to B$, A being the donor and B the acceptor. This indicates, in a convenient way, the origin of the electrons making up the bond.

The second shows an $A \to B$ bond as $A^{\oplus}—B^{\ominus}$. This method indicates the electrical charges which develop on atoms A and B as a result of dative bond formation. A, the donor, develops a positive charge by partly transferring two electrons to B; B, the acceptor, develops a corresponding negative charge. On this basis, the dative bond can be regarded as a covalent bond with a certain amount of ionic character, and the term co-ionic, instead of dative, is intended to describe this state of affairs. Other terms which have been used are co-ordinate bond, semi-polar bond or semi-polar double bond. The use of dative is preferred because it was the term finally used by Sidgwick, who first developed the use of this type of bond.

The following structural formulae show the various ways in which dative bonds can occur in typical compounds.

(a) *Nitromethane*, $CH_3.NO_2$.

(b) *Isocyanomethane*, $CH_3.NC$,

(c) *The ammonium ion*, NH$_4^+$,

$$\left\{ \begin{array}{c} H \\ H : N : H \\ H \end{array} \right\}^+$$

(d) *Aluminium chloride*, Al$_2$Cl$_6$.

```
   Cl       Cl       Cl
     \     / \     /
      Al       Al
     /     \ /     \
   Cl       Cl       Cl
```

5 Other types of chemical bonding Ionic, covalent and dative bonding cannot satisfactorily account for all the cohesive forces which seem to be found in chemical substances. Other important types of chemical bonding are mentioned here and discussed later.

(a) *The metallic bond* (page 209). Metals have very distinctive properties and to account for these, particularly for the electrical conductivity, the idea of a special metallic bond is necessary.

(b) *The hydrogen bond* (page 141). A hydrogen atom can often link together two electronegative atoms, and the resulting hydrogen bond, though considerably weaker than a covalent bond, is of importance.

(c) *van der Waals' forces* (page 154). These forces hold molecules together in the liquid or solid state, but provide comparatively weak binding.

6 Noble gas compounds For many years it was thought that the stability of the arrangement of electrons in noble gas atoms was such that they could not participate in chemical combination. This, indeed, was one of the important foundation-stones of early valency theory (page 32). Such ideas were dispelled, however, in 1962, when Bartlett made an orange-yellow solid, with a formula XePtF$_6$, by treating xenon at room temperature with platinum hexafluoride, a very reactive gas. He was led into doing this by his prior discovery that platinum hexafluoride reacted with oxygen to form an ionic compound, [O$_2$]$^+$[PtF$_6$]$^-$, and he argued that platinum hexafluoride might oxidise Xe to Xe$^+$ if it could oxidise O$_2$ to O$_2^+$.

This first breakthrough soon led to the isolation of other compounds of xenon. The fluorides, XeF$_2$, XeF$_4$ and XeF$_6$, are the stablest and most readily made products; they are white solids. Oxyfluorides, XeOF$_4$ and XeO$_2$F$_2$, have also been made, and an oxide, XeO$_3$ exists, but it is an explosive white solid. The di- and tetra-fluorides of krypton, and a fluoride of radon, have also been made, but no compounds of argon, helium and neon have yet been reported.

It is probably significant that xenon forms compounds the most easily of all the noble gases, and that it does so with the highly electronegative elements, oxygen and fluorine. Both these points suggest (page 42) that the bonding in the noble gas compounds is essentially ionic, e.g. Xe$^+$[PtF$_6$]$^-$, Xe^{2+}(F$^-$)$_2$ but the detailed nature of the bonding is still to be elucidated.

5 The Formation of Ionic Bonds

1 Stable ionic structures The only known negatively charged ions (anions) have an inert gas structure, the commonest simple anions being

N^{3-} 2.8 O^{2-} 2.8 H^- 2
P^{3-} 2.8.8 S^{2-} 2.8.8 F^- 2.8
                                  Cl^- 2.8.8
                                  Br^- 2.8.18.8
                                  I^- 2.8.18.18.8

with structures as shown.

Cations can, however, exist with much more varied structures as described below.

(*a*) *Cations with noble gas structures.* Of the various types of cation, those with noble gas structure are easily the most stable and, as a result, have the smallest tendency to form complex ions. The simplest noble gas type cations can be related to their noble gas structures as follows:

He structure, 2 Li^+ Be^{2+}
Ne structure, 2.8 Na^+ Mg^{2+} Al^{3+}
Ar structure, 2.8.8 K^+ Ca^{2+} Sc^{3+}
Kr structure, 2.8.18.8 Rb^+ Sr^{2+} Y^{3+} Zr^{4+}
Xe structure, 2.8.18.18.8 Cs^+ Ba^{2+} La^{3+} Ce^{4+}

The elements forming these noble gas-type cations are to be found in groups 1A, 2A, 3A and 4A of the periodic table (page 29).

(*b*) *Cations with 18-electron group structures.* Many metals in the B sub-groups of the periodic table form cations which have not got a noble gas structure, but which have 18 electrons in the outermost orbit. Examples of some of the commoner 18-electron group cations, together with their electronic structures are given below:

2.8.18 Cu^+ Zn^{2+} Ga^{3+}
2.8.18.18 Ag^+ Cd^{2+} In^{3+} Sn^{4+}
2.8.18.32.18 Au^+ Hg^{2+} Tl^{3+} Pb^{4+}

These ions are well known, but they are not so stable as ions with a noble gas structure. This explains the greater ease of formation of, for instance, a calcium ion, Ca^{2+}, as compared with a zinc ion, Zn^{2+}. It also explains why Cu^+ ions more readily form complex ions than K^+ ions.

The arrangement of 18 electrons in the outermost orbit is not very stable in elements in which the group of 18 electrons has only just filled up. In such cases one or more of the 18 electrons can be lost quite easily to form an ion with a greater charge. Copper, silver and gold, for instance, all form other than monovalent ions as shown below:

$$\begin{array}{lll} \text{Cu} \quad 2.8.18.1 & \text{Ag} \quad 2.8.18.18.1 & \text{Au} \quad 2.8.18.32.18.1 \\ \text{Cu}^+ \quad 2.8.18 & \text{Ag}^+ \quad \underline{2.8.18.18} & \text{Au}^+ \quad 2.8.18.32.18 \\ \text{Cu}^{2+} \quad \underline{2.8.17} & \text{Ag}^{2+} \quad 2.8.18.17 & \text{Au}^{3+} \quad \underline{2.8.18.32.16} \end{array}$$

The more stable ion in each case is underlined.

The loss of electrons from the 18-group must be due to the charge on the nucleus not being high enough to hold the 18-group firmly, for in the elements with the next highest atomic numbers, i.e. the next stronger nuclear attractions, none of the 18-group electrons can be lost. The result is that zinc, cadmium and mercury can form divalent, but not trivalent, ions:

$$\begin{array}{lll} \text{Zn} \quad 2.8.18.2 & \text{Cd} \quad 2.8.18.18.2 & \text{Hg} \quad 2.8.18.32.18.2 \\ \text{Zn}^{2+} \quad 2.8.18 & \text{Cd}^{2+} \quad 2.8.18.18 & \text{Hg}^{2+} \quad 2.8.18.32.18 \end{array}$$

(*c*) *Cations of transitional elements.* In ions formed from transitional elements, any number of electrons, from 9 to 17, can occur in the outermost orbit, and the ions do not have a noble gas structure. The ions of transitional elements readily form complexes, and there is such a small difference between the stability of two or more alternative structures that the elements commonly have variable valency and form two or more simple ions. Iron, cobalt and nickel, for example, all form both divalent and trivalent ions:

$$\begin{array}{lll} \text{Fe} \quad 2.8.14.2 & \text{Co} \quad 2.8.15.2 & \text{Ni} \quad 2.8.16.2 \\ \text{Fe}^{2+} \quad 2.8.14 & \text{Co}^{2+} \quad 2.8.15 & \text{Ni}^{2+} \quad 2.8.16 \\ \text{Fe}^{3+} \quad 2.8.13 & \text{Co}^{3+} \quad 2.8.14 & \text{Ni}^{3+} \quad 2.8.15 \end{array}$$

In this series the divalent ion becomes more stable in passing from iron to nickel. It is reasonable to assume that this is due to the greater nuclear charge on the nickel atom which can hold all the electrons more firmly.

2 The inert pair effect Some of the heavier sub-group B elements, which would be expected to form only ions with an 18-electron group, do, in fact, form other ions too. These ions have a charge of two units less than that of the expected ion, i.e. two electrons do not play their full part in ion formation. Such electrons are known as an inert pair; they are *s* electrons, and the extra stability associated with a full *s* level in an atom is shown by the ionisation energy values discussed on page 44.

The following well-established facts are all explained on this idea of an inert pair of electrons:

(a) *The existence of* Tl^+ *and* In^+ *ions.* Both thallium and indium form trivalent ions (page 38) but they also form monovalent ions,

    Tl 2.8.18.32.18.3   In 2.8.18.18.3
    Tl^{3+} 2.8.18.32.18 In^{3+} 2.8.18.18
    Tl^+ 2.8.18.32.18.2 In^+ 2.8.18.18.2

Compounds of thallium(I) are more stable than those of thallium (III), i.e. thallium(III) salts are oxidising agents. In^+ ions, however, are immediately converted into In^{3+} ions and the metal in the presence of water. Gallium forms no monovalent ions.

The Tl^+ ion has the same electronic arrangement as the Pb^{2+} ion (see (b) below) and this explains the likenesses between certain thallium(I) and lead(II) compounds. Thus thalium(I) chloride, bromide, iodide, sulphate and sulphide are only slightly soluble like the corresponding lead(II) compounds. The similarities between thallium(I) compounds and compounds of the alkali metals are due to the fact that the thallium(I) ion has about the same size, and the same charge, as the ions of the alkali metals (page 53).

(b) *The existence of* Sn^{2+} *and* Pb^{2+} *ions.* Tin and lead both form the expected four-valent ions, but they also form divalent ions,

    Sn 2.8.18.18.4   Pb 2.8.18.32.18.4
    Sn^{4+} 2.8.18.18 Pb^{4+} 2.8.18.32.18
    Sn^{2+} 2.8.18.18.2 Pb^{2+} 2.8.18.32.18.2

The effect of the inert pair is more marked in the heavier element, i.e. lead, and the Sn^{4+} ion is more stable than the Sn^{2+} ion, whereas the Pb^{2+} ion is more stable than the Pb^{4+}. This explains why tin(II) oxide and tin(II) chloride are reducing agents, whereas lead(IV) oxide, PbO_2, and lead tetrachloride are oxidising agents.

Moreover, lead tetrabromide and lead tetraiodide do not exist but the corresponding compounds of tin can be obtained.

(c) *The existence of* Sb^{3+} *and* Bi^{3+} *ions.* Both antimony and bismuth form trivalent ions,

    Sb 2.8.18.18.5   Bi 2.8.18.38.18.5
    Sb^{3+} 2.8.18.18.2 Bi^{3+} 2.8.18.32.18.2

and give definite trivalent salts. As with tin and lead the effect of the inert pair is more marked in the heavier element, i.e. bismuth. Bismuth, for instance, unlike phosphorus, arsenic, and antimony does not form a pentachloride.

The effect of the inert pair in arsenic is apparent only in a few complex compounds.

The fact that mercury vapour is monatomic, like the inert gases, points to mercury behaving as Hg, 2.8.18.32.18 and not as Hg, 2.8.18.32.18.2, i.e. a pair of electrons seems to be inert. Further less simple evidence leads to the conclusion that the inert pair effect shows itself in the lower half of the table given below, i.e. in the heavier B sub-group elements:

Be	B	C	N	O	F
Mg	Al	Si	P	S	Cl
Zn	Ga	Ge	As	Se	Br
Cd	In	Sn	Sb	Te	I
Hg	Tl	Pb	Bi	—	—

The effect in any one vertical series becomes more marked in passing down the series.

3 Limitations to the formation of ions Whether or not an atom will form an ion depends to some extent on the stability of the ionic structure which it might form. As ionisation depends on the gaining (anion formation) or losing (cation formation) of electrons, however, both the size of the atom and its atomic number are also important. The size of the atom determines the distance of the valency electrons from the nucleus. The atomic number determines the positive charge on the nucleus, and it is this charge which holds the electrons in position.

The farther an electron is from the nucleus, the less firmly is it held and the more easily can it be lost. Thus in the group lithium, sodium, potassium, rubidium and caesium, the last atom, which is the largest, forms an ion most easily. The loss of an electron from a caesium atom is so easy that caesium shows a very strong photoelectric effect (page 67). Similarly, barium ionises more readily than strontium, calcium or magnesium.

Once an electron has been lost from an atom, the remaining ones are held more firmly and are not lost so easily. It is in this way that the formation of cations is limited to those with a charge of four units, and such highly charged ions are rare and only formed by large atoms. Tin and lead do form quadrivalent ions, but the smaller atoms of carbon and silicon do not. Similarly, aluminium forms a trivalent ion, but the smaller boron atom does not.

In the formation of anions, the positive charge on the nucleus of an atom may be able to hold one extra electron, and can sometimes hold two or three, but three is the limit. The smaller the anion, the more easily can the nuclear charge hold the extra electrons, for, in a small ion, they are nearer to the nucleus than in a larger one. As a result, simple anions are limited to those of hydrogen (in salt-like hydrides such as lithium hydride, LiH), the halogens, oxygen, sulphur, selenium and tellurium, and nitrogen and phosphorus (in some nitrides and phosphides). In the halogen series, fluorine, with the smallest atom, most easily forms an anion. Thus mercury(II), aluminium and tin(IV) fluorides are ionic compounds, whereas the corresponding chlorides are covalent.

4 Fajans's rules From the general considerations outlined in the preceding section, it is possible to summarise the conditions favouring the formation of an ion. An ion will be formed most easily:

(*a*) if the electronic structure of the ion is stable,

(*b*) if the charge on the ion is small, and

(*c*) if the atom from which the ion is formed is small for an anion, or large for a cation.

These rules, in a different form, were first suggested by Fajans, in 1924, and are usually known as Fajans's rules.

Fajans originally expressed the rules in terms of the concept of atomic volume, i.e. the relative atomic mass of an element divided by its density. This gives an approximate measure of the size of an atom of an element and it follows from what has been said that an ion will be formed most easily if the element from which the ion is formed has a low atomic volume for an anion and a large atomic volume for a cation. If the conditions prevailing in any particular case do not favour the formation of ions then a covalent bond will probably be formed rather than an ionic bond.

Some general applications of Fajans's rules are outlined below.

(*a*) *Ease of formation of ions.* For ions which have a noble gas structure the predominant factors in their formation are the ionic size and the ionic charge. The effect of these two factors on the formation of common cations and anions can be summarised as follows:

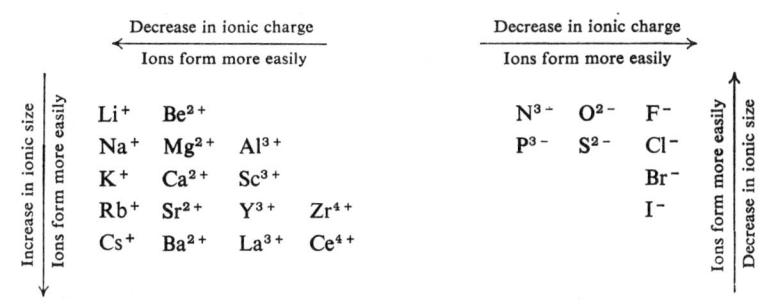

	Decrease in ionic charge → Ions form more easily				Decrease in ionic charge → Ions form more easily			
Increase in ionic size ↓ / Ions form more easily ↓	Li⁺	Be²⁺			N³⁻	O²⁻	F⁻	Ions form more easily ↑ / Decrease in ionic size ↑
	Na⁺	Mg²⁺	Al³⁺		P³⁻	S²⁻	Cl⁻	
	K⁺	Ca²⁺	Sc³⁺				Br⁻	
	Rb⁺	Sr²⁺	Y³⁺	Zr⁴⁺			I⁻	
	Cs⁺	Ba²⁺	La³⁺	Ce⁴⁺				

(*b*) *Physical properties of chlorides.* The way in which the size and charge of the cation affects the types of chloride formed is well illustrated by the figures given in Table 1 for the melting points, boiling points and equivalent conductivities (in the fused state) of the chlorides of typical and A sub-group elements. The electronic structures of all the cations which might be considered as being involved is a noble gas structure so that no effect due to differences in the stability of the ions arises.

These chlorides clearly fall into two groups. Those beneath the diagonal line are electrolytes and probably contain ionic bonds, whilst those above the line are non-electrolytes and probably contain covalent bonds. It is not, however, justifiable to be dogmatic about this on the evidence given, for the behaviour of a substance in the molten state does not necessarily have any significance as to its structure as a solid. Nevertheless the diagonal demarcation line does seem to be related to the two tendencies indicated in the table, i.e. an increase in ionic charge along a horizontal period coupled with an increase in ionic size in going down a vertical group.

(*c*) *Diagonal relationships.* The so-called diagonal relationships in the periodic table shown below,

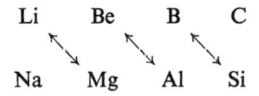

$$\begin{array}{cccc} \text{Li} & \text{Be} & \text{B} & \text{C} \\ \searrow & \searrow & \searrow & \\ \text{Na} & \text{Mg} & \text{Al} & \text{Si} \end{array}$$

are due to the comparative ease of ionisation of the elements joined by arrows. Any increase in ease of ionisation in passing down a group is counteracted by a decrease in the ease of ionisation in passing from left to right. What similarities there are between, say, lithium and magnesium are, therefore, due to the fact that they both form ions with the same ease.

TABLE 1

The Melting Points (a), Boiling Points (b), and Equivalent Conductivities in the Fused State (c), of Chlorides of Typical and A Sub-group Elements

Increasing ionic charge →

Increasing ionic size ↓

HCl (a) −114° (b) −85° (c) <10⁻⁶					
LiCl (a) 606° (b) 1,382° (c) 166	BeCl₂ 404° 488° 0·086	BCl₃ −107° 12·5° 0	CCl₄ −23° 76° 0	NCl₅ Not formed	OCl₆ Not formed
NaCl (a) 800° (b) 1,440° (c) 133	MgCl₂ 715° (1,410°) 29	AlCl₃ — 183° 15×10⁻⁶	SiCl₄ −70° 57° 0	PCl₅ 148° Decomps. 0	SCl₆ Not formed
KCl (a) 768° (b) 1,415° (c) 103	CaCl₂ 774° (1,600°) 52	ScCl₃ — (1,000°) 15	TiCl₄ −23° 136° 0	VCl₅ Not formed	CrCl₆ Not formed
RbCl (a) 717° (b) 1,383° (c) 78	SrCl₂ 870° (1,250°) 56	YCl₃ — (700°) 9·5	ZrCl₄ 437° (25 atms.) Sublimes —	NbCl₅ 194° 241° 2·2×10⁻⁷*	MoCl₆ Not formed
CsCl (a) 640° (b) 1,303° (c) 67	BaCl₂ 955° (1,800°) 65	LaCl₃ — (1,000°) 29	HfCl₄ 432° — —	TaCl₅ 211° 242° 3×10⁻⁷*	WCl₆ 275° 347° 2×10⁻⁶*
			ThCl₄ (a) 765° (b) 922° (c) 16	* Conductivities.	

5 Ionisation energy or ionisation potential A direct measure of the ease with which an atom or an ion can lose an electron is provided by the ionisation energy or the ionisation potential of the atom or ion. This is the energy required to withdraw an electron against the attraction of the nuclear charge. In other words, it is the heat of the reaction of the change

$$\text{Atom } (A) + \text{energy} \rightarrow \text{Cation } (A^+) + 1 \text{ electron}$$

This, in fact, is known as the first ionisation energy or potential of the atom for it refers to the loss of the first or outermost electron. Other ionisation energy values refer to the loss of further electrons. Thus the second ionisation energy for A is the same as the first ionisation energy for A^+, and so on.

$$A \xrightarrow{-1e} A^+ \xrightarrow{-1e} A^{++}$$

Ionisation energies can be measured spectroscopically (page 14), or by electron bombardment, or by measuring the current passing through a discharge tube, containing gas or vapour of the element under consideration, as the applied voltage is gradually increased. At certain voltages there are marked rises in the current passing and these correspond to the points at which atoms of the element concerned lose one, two, three or more electrons. The lost electrons enhance the current.

Ionisation energies can be expressed in kJ mol^{-1} or kcal mol^{-1} or in electron volts; the relationship between these two units is explained on page 14. The experimental figures illustrate quantitatively some of the effects which have already been mentioned qualitatively.

(a) *Effect of size of ion.* The lower the ionisation energy of an atom or ion the more easily will that atom or ion lose an electron. The following horizontal series show how the ionisation energy decreases, i.e. the ease of ionisation increases, as the size of comparable ions increases. The values are given in kJ mol^{-1}.

Li → Li$^+$	Na → Na$^+$	K → K$^+$	Rb → Rb$^+$	Cs → Cs$^+$
521	497·9	422·6	401·7	380·7
Be → Be^{2+}	Mg → Mg^{2+}	Ca → Ca^{2+}	Sr → Sr^{2+}	Ba → Ba^{2+}
2657	2176	1724	1602	1460
B → B^{3+}	Al → Al^{3+}	Sc → Sc^{3+}	Y → Y^{3+}	La → La^{3+}
5623	5099	4255	3775	3493

(b) *Effect of charge on ion.* The above figures, taken in vertical columns, show that it is easier to form an ion with a low charge, e.g. Na$^+$, than one with a higher charge, e.g. Al^{3+}. This is because it becomes progressively more difficult to remove electrons from an atom after the first one. Once one has been removed, the remainder are more tightly held by the nuclear attraction and the effect is still greater when two or more have been removed. Thus the first, second, third, etc. ionisation energies for any one atom increase, as in the following series:

Ionisation energies in kJ mol^{-1}

	1st $M \to M^+$	2nd $M^+ \to M^{2+}$	3rd $M^{2+} \to M^{3+}$	4th $M^{3+} \to M^{4+}$	5th $M^{4+} \to M^{5+}$	6th $M^{5+} \to M^{6+}$	7th $M^{6+} \to M^{7+}$
H	1 311						
He	2 372	5 249					
Li	521	7 297					
Be	899	1 758	14 850				
B	801	2 428	3 658	25 020			
C	1 086	2 353	4 618	6 512	37 830		
N	1 400	2 855	4 577	7 473	9 448		
O	1 310	3 388	5 297	7 450	10 990	13 320	
F	1 680	3 375	6 045	8 409	11 030	15 160	17 850

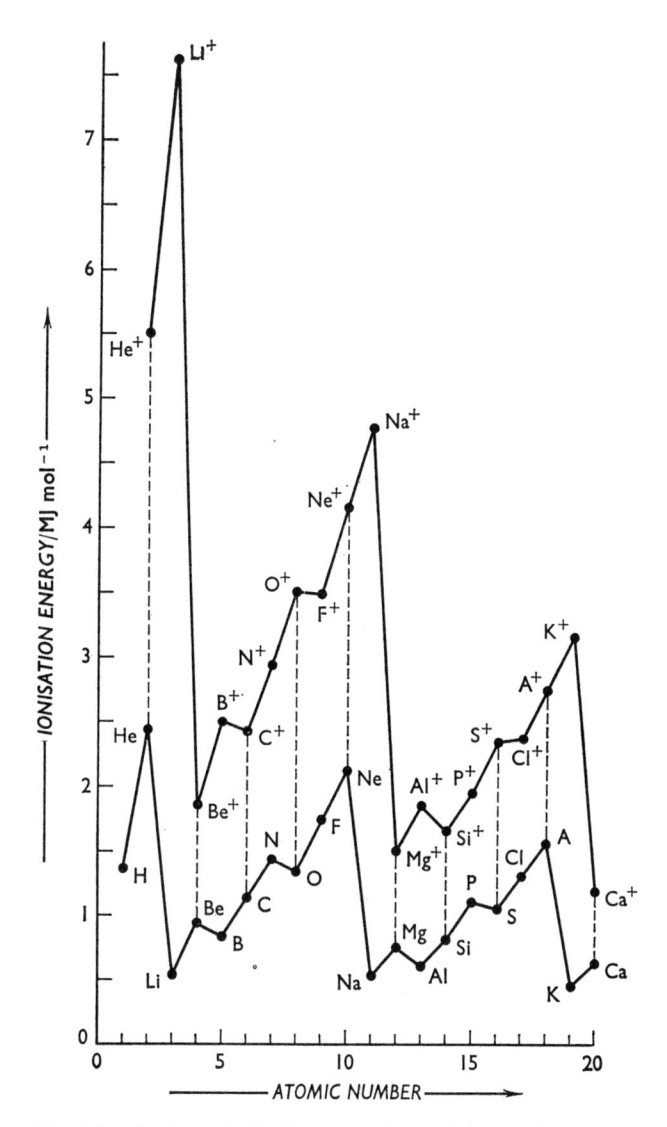

Fig. 13. Graph of ionisation energies against atomic number. The lower line shows the first ionisation energy. The upper line shows the second ionisation energy of the atom or the first ionisation energy of the ion, X^+.

(c) *Effect of arrangement of electrons.* Plotting first and second ionisation energies against atomic number for the first twenty elements gives the result shown in Fig. 13. Similar curves are obtained for the third and fourth ionisation energies, and, also, for elements of atomic number greater than twenty.

The curves show that atoms or ions with a noble gas arrangement of electrons, i.e. He, Ne, Ar, Li$^+$, Na$^+$ and K$^+$, occupy the maxima. These atoms and ions have the highest ionisation energies because they have the most stable arrangement of electrons.

Elements or ions with one electron more than a noble gas structure, i.e. Li, Na, K, Be$^+$, Mg$^+$ and Ca$^+$, have the lowest ionisation energies.

Irregularities occur with Be and B$^+$, N and O$^+$, Mg and Al$^+$ and P and S$^+$, all these atoms or ions having a slightly higher ionisation energy than would be expected. This is attributed to the extra stability of a full s level in Be, B$^+$, Mg and Al$^+$, and to the extra stability of a half-full p level in N, O$^+$, P and S$^+$.

6 Electron affinity The ionisation energy gives a measure of the ease with which an atom or an ion can lose electrons; it is, therefore, concerned with the formation of cations. The corresponding quantity for anion formation is known as the electron affinity. This is the energy given out when an extra electron is taken up by an atom or ion. It is a measure of the heat of reaction for the change

$$\text{Atom}(A) + 1 \text{ electron} \rightarrow \text{Anion}(A^-) \quad \Delta H = \text{Electron affinity}$$

and may also be regarded as the ionisation energy of the anion, for the energy given out when an electron is added to an atom is clearly equal to that required to remove an electron from the anion.

Electron affinities cannot be measured very easily. Spectroscopic methods can be used, and one main method depends on a study of the change of equilibrium of the above reaction with temperature. Electron affinities can also be estimated by applying the principle of the Born-Haber cycle (page 62). Some values, in kJ mol^{-1}, are given below:

H → H$^-$			
−72			

F → F$^-$	Cl → Cl$^-$	Br → Br$^-$	I → I$^-$
−333	−364	−342	−295

O → O^{2-}	S → S^{2-}
+791	+649

Once an atom has had one extra electron added to it it becomes negatively charged and the addition of more electrons is opposed.

That is why there are positive values for the oxide and sulphide ions, and why the halide ions are much more readily formed.

7 Electrode potentials Both ionisation energies and electron affinities give a measure of the ease with which atoms will form ions, but the information is, sometimes, only of theoretical interest for the values of ionisation energies and electron affinities refer to electron-loss or electron-gain *by isolated gaseous atoms*. For changes occurring in solution, electrode potentials give the best measure of the relative ease of ionisation.

The standard electrode potential of an element is the potential difference between the element and a solution containing the ions formed from the element in molar concentration, i.e. 1 mol dm^{-3}. The electrode potential of copper, for example, is the potential difference between a rod of copper and an M solution of copper(II) sulphate(VI) in which it is immersed. The potential difference is set up because the tendency for the copper to form Cu^{2+} ions (known as the electrolytic solution pressure) is not equally balanced by the tendency for Cu^{2+} ions to deposit on the copper from the solution (the deposition pressure).

For a copper rod in M copper(II) sulphate solution the deposition pressure is greater than the electrolytic solution pressure, i.e. the change

$$Cu^{2+} (aq) + 2e \rightarrow Cu(s)$$

predominates. As the electrons necessary for this change come from the copper rod it assumes a positive charge in relation to the solution. The copper is said to have a positive electrode potential. To enable actual measurements to be made the electrode potential of hydrogen is arbitrarily taken as zero, and other values are measured on this scale. That for copper is +0·34 volt.

Zinc, in comparison, has a negative electrode potential of −0·76. This is because the change

$$Zn(s) \rightarrow Zn^{2+} (aq) + 2e$$

predominates when a zinc rod is immersed in an M solution of Zn^{2+} ions.

In simple language, zinc forms ions much more readily than copper does. And, in general, an element with a high negative electrode potential is one which readily loses electrons and forms positive ions, whilst a large positive electrode potential indicates a readiness to form negative ions by gain of electrons.

The resulting list of electrode potentials given in Table 2 is known as the electrochemical series or the reactivity series.

TABLE 2

The Electrochemical Series
(Values in volt)

Cs → Cs$^+$	−2·93	Al → Al^{3+}	−1·28	H → H$^+$	0·00
Rb → Rb$^+$	−2·92	Zn → Zn^{2+}	−0·76	Bi → Bi^{3+}	+0·20
K → K$^+$	−2·92	Cr → Cr^{2+}	−0·56	Cu → Cu^{2+}	+0·34
Ba → Ba^{2+}	−2·84	Fe → Fe^{2+}	−0·44	I → I$^-$	+0·58
Sr → Sr^{2+}	−2·74	Cd → Cd^{2+}	−0·40	Hg → Hg^{2+}	+0·80
Na → Na$^+$	−2·71	Tl → Tl$^+$	−0·34	Ag → Ag$^+$	+0·80
Ca → Ca^{2+}	−2·56	Co → Co^{2+}	−0·29	Br → Br$^-$	+1·08
Li → Li$^+$	−2·09	Ni → Ni^{2+}	−0·22	Cl → Cl$^-$	+1·36
Mg → Mg^{2+}	−1·55	Sn → Sn^{2+}	−0·14	Au → Au^{3+}	+1·37
		Pb → Pb^{2+}	−0·12	F → F$^-$	+2·80

6 Characteristics of Ionic Compounds

1 General characteristics Many compounds can be satisfactorily formulated with ionic bonds, and they tend to have certain characteristic properties which may be summarised as follows:

(*a*) Individual molecules of ionic compounds do not exist because the compounds are made up of an interlocking structure of ions. The ions are held together by strong electrostatic forces within an ionic crystal, the arrangement of ions within such a crystal, i.e. the crystal structure, being mainly determined by the charges and sizes of the ions concerned (page 55). The formula of an ionic compound simply shows the relative numbers of each ion present.

(*b*) Ionic compounds are invariably electrolytes, for, in the presence of an ionising solvent, such as water, the forces between the ions are so greatly reduced that the ions 'fall apart', and the free ions, in the resulting solution, are able to move under the influence of an electrical field as in electrolysis.

(*c*) Ionic compounds are often hard solids because the inter-ionic forces within an ionic crystal are usually strong.

(*d*) Ionic compounds generally have high melting points because a lot of thermal energy is required to break down the inter-ionic forces and form a liquid. Once an ionic compound has been melted, the melt can undergo electrolysis because it contains free ions. A high melting point is associated, too, with a high boiling point for ionic compounds.

(*e*) Ionic compounds are commonly soluble in water, or other ionising solvents, and insoluble in benzene or other organic solvents.

These are very broad generalisations, and they must be treated as such; it is not difficult to find exceptions.

THE SIZE OF IONS

2 Ionic radii X-ray analysis of crystals, and other methods, give values for the equilibrium distance between the nuclei of two adjacent ions in an ionic crystal. The values are very small and are generally expressed in nanometres (nm), 1 nm being equal to 10^{-9} m. Ångstrom units (p. 7) may also be used. The internuclear distances for the halides of sodium and potassium are given below in nm

	KF	KCl	KBr	KI
	0·266	0·314	0·329	0·353
	NaF	NaCl	NaBr	NaI
	0·231	0·281	0·298	0·323
Difference	0·035	0·033	0·031	0·030

If the ions in a crystal are regarded as spheres, the internuclear distance between two ions will be made up of the sum of the ionic radii of the two ions (Fig. 14). Moreover, the constancy of the differ-

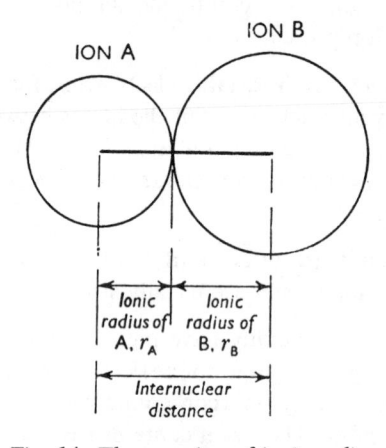

Fig. 14. The meaning of ionic radius

ence between the internuclear distances of sodium and potassium halides, as shown, indicates that the ionic radii of anions and cations must be reasonably constant in a series of such compounds.

The actual internuclear distance does not give a value for an ionic radius until any one ionic radius is decided by some other method. Different, rather arbitrary, methods have been adopted in order to fix the absolute value of the ionic radius of an ion. Pauling's method is the most widely accepted. He considered the situation in a bond between two iso-electronic ions, such as K^+ and Cl^-. Both these ions have an equal number of electrons, but they have different nuclear charges, and Pauling argued that the internuclear distance between the two ions would be divided in the inverse ratio of their effective nuclear charges. Taking into account the screening effect of the electrons in order to calculate the effective nuclear charge. Pauling accepted radii values for K^+ and Cl^- of 0·133 nm and 0·181 nm. Other values were then readily obtainable from measured internuclear distances, and are summarised below in nm:

Li^+			Be^{2+}			
0·060			0·031			

Na^+			Mg^{2+}		Al^{3+}	
0·095			0·065		0·050	

K^+	Cu^+	Ca^{2+}	Zn^{2+}	Sc^{3+}	Ga^{3+}	Ti^{4+}
0·133	0·096	0·099	0·074	0·081	0·062	0·068

Rb^+	Ag^+	Sr^{2+}	Cd^{2+}	Y^{3+}	In^{3+}	Zr^{4+}	Sn^{4+}
0·148	0·126	0·113	0·097	0·093	0·081	0·080	0·071

Cs^+	Au^+	Ba^{2+}	Hg^{2+}	La^{3+}	Tl^{3+}		Pb^{4+}
0·169	0·137	0·135	0·110	0·115	0·095		0·084

F^-	Cl^-	Br^-	I^-
0·136	0·181	0·195	0·216

O^{2-}	S^{2-}	Se^{2-}	Te^{2-}
0·140	0·184	0·198	0·221

Ionic Radii in nm

3 Trends in values of ionic radii A study of the numerical values given leads to the following pertinent conclusions, most of which simply serve to illustrate, numerically, general effects which have already been described:

(*a*) The ions of elements in any one group of the periodic table increase in size as the relative atomic mass of the element increases.

(*b*) For a series of ions with the same arrangement of electrons the size of the ion decreases as the nuclear charge, i.e. the atomic number, increases, for, as the nuclear charge increases, the electrons are attracted more strongly and drawn inwards. This effect is observable in the series O^{2-}, F^-, Na^+, Mg^{2+}, Al^{3+} (all with a 2.8 structure), S^{2-}, Cl^-, K^+, Ca^{2+}, Sc^{3+} (all with a 2.8.8 structure), or Cu^+, Zn^{2+}, Ga^{3+} (all with a 2.8.18 structure).

(*c*) When an element forms two positively charged ions the ion with the lower charge is larger than the more highly charged ion. This effect is to be expected for the ion with the higher charge has fewer extra-nuclear electrons, and these are held more tightly, than the ion of lower charge. Thus:

Tl^+	Pb^{2+}	Mn^{2+}	Fe^{2+}
0·144	0·121	0·080	0·075
Tl^{3+}	Pb^{4+}	Mn^{3+}	Fe^{3+}
0·095	0·084	0·062	0·060

(*d*) There is a decrease in ionic radius as the atomic number of the lanthanides or rare-earths increases: this effect is known as the *lanthanide contraction*, and is illustrated by the figures, in nm, given below:

La	Ce	Pr	Nd	Pm	Sm	Eu	Gd
0·115	0·111	0·109	0·108	0·106	0·104	0·103	0·102
Tb	Dy	Ho	Er	Tm	Yb	Lu	
0·100	0·099	0·097	0·096	0·095	0·094	0·093	

In passing along the series there is an increase in nuclear charge and an increase in the number of electrons. But the added electrons are in the 4*f* shell, which is so far from the outside of the atom that extra electrons placed in it have little effect on the size.

A similar decrease in ionic radii with increase in atomic number is found in the actinides and in the transitional elements.

4 Effect of co-ordination number on ionic radii The ionic radii given on page 53 are based on the assumption that an ion may be regarded as a sphere. This is not really accurate as, on the one hand, the arrangement of electrons is rather diffuse, which makes it difficult to define the perimeter of an ion, and, on the other, two adjacent ions interact and cause deformation.

There are no grounds for assuming that the deformation of an ion is the same in one type of crystal arrangement as in another, and it is, therefore, necessary to allot slightly different ionic radii to ions when such ions occur in different crystal structures. The values given on page 53 are those for ions taking part in a crystal structure with a co-ordination number of 6 (page 55).

Ammonium chloride crystallises with the sodium chloride structure (co-ordination number, 6) above 184·3°C and with the caesium chloride structure (co-ordination number, 8) below that temperature, and there is found to be an interionic distance about 3 per cent greater in the latter arrangement. On these lines conversion factors can be worked out for converting the ionic radii for sixfold co-ordination to ionic radii for any other co-ordination. Thus for co-ordination number from 6 to 8 the factor is 1·03, and from 6 to 4 it is 0·94. It will be seen that the changes are not very great.

STRUCTURES OF IONIC CRYSTALS

5 Co-ordination number In a crystal made up of A^+ and B^- ions, the number of B^- ions immediately surrounding an A^+ ion is known as the co-ordination number of A^+, whilst the number of A^+ ions immediately surrounding a B^- ion is the co-ordination number of B^-. In a binary compound, AB, the co-ordination numbers of A^+ and B^- must be equal, but in AB_2 compounds the co-ordination number of A^+ will be twice that of B^-.

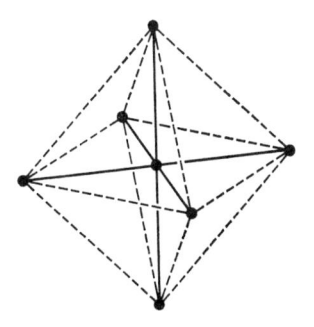

Fig. 15. Octahedral arrangement of six ions around a central ion

The co-ordination number which will give the most stable structure for any pair of ions, A^+ and B^-, is dependent on the ratio of the radius of A^+ (r^+) to that of B^- (r^-), i.e. on r^+/r^-.

Consider, for example, a structure in which a central A^+ ion is surrounded, octahedrally, by six B^- ions; the co-ordination number will be 6. Figure 15 shows the general arrangement in such a structure but indicates only the nuclei of the ions concerned and not their actual relative sizes. Figure 16 shows the possible relative sizes of A^+ and B^- in the limiting case. There will be B^- ions both directly above and below the central A^+ ion, but these have been omitted for the sake of clarity. In this limiting case, the B^- ions are in contact with each other as well as with A^+. Such a state of affairs can only exist when $r^-/(r^+ + r^-)$ is equal to cos 45°, i.e. when r^+/r^- is equal to 0·414.

If different ions, with larger r^- and/or smaller r^+ values, are concerned, a stable octahedral arrangement would be much less likely for, as shown in Fig. 17, the B^- ions would be in contact with each other, and repelling each other, without being in contact with, and being attracted by, A^+. Such ions would rearrange into a tetrahedral structure with a co-ordination number of 4. In simple language, the cation has become too small, or the anion too large, for the central cation to be surrounded by six anions; instead, it is surrounded by four.

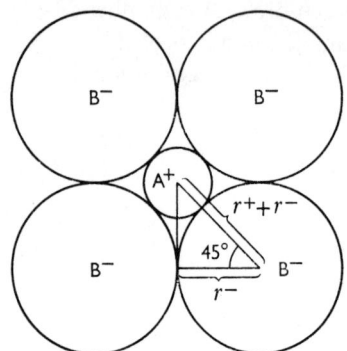

Fig. 16. Showing the limiting r^+/r^- value for an octahedral arrangement

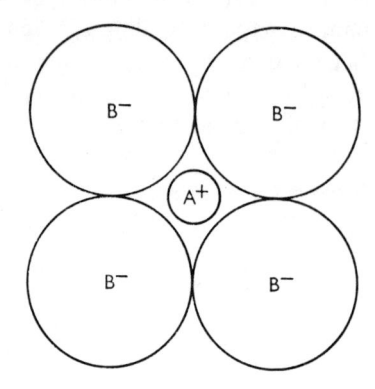

Fig. 17. The unstable octahedral arrangement when r^+/r^- becomes less than 0.414

If the radius ratio, r^+/r^-, becomes smaller still, the central cation may only be able to be surrounded by three anions in a coplanar, triangular arrangement with a co-ordination number of 3. Such an arrangement is shown in Fig. 18, from which it can be seen that the limiting radius ratio for such a structure occurs when $r^-/(r^+ + r^-)$ is equal to cos 30°, i.e. when r^+/r^- is equal to 0·155.

By similar arguments and calculations the limiting radius ratio values for various co-ordination numbers can be summarised as follows:

Structure	Co-ordination number	Limiting radius ratio
Linear	2	0–0·155
Triangular	3	0·155–0·225
Tetrahedral	4	0·225–0·414
Octahedral	6	0·414–0·732
Cubic	8	0·732–1

Co-ordination numbers of 5, 7, 9, 10 and 11 do not occur because of the impossibility of balancing the electrical charges of the ions concerned. When the radius ratio becomes equal to 1, ions of the same size are making up the crystal. Such a state of affairs is to be found in crystals of metals (page 222).

Radius ratio values are useful, but they do not apply absolutely accurately. This is only to be expected for the whole conception of ionic radius, based as it is on the idea that an ion is always spherical, is only an approximation.

6 The sodium chloride crystal structure The radius ratio for Na^+ and Cl^- ions is 0·095/0·181 or 0·525 so that an octahedral structure

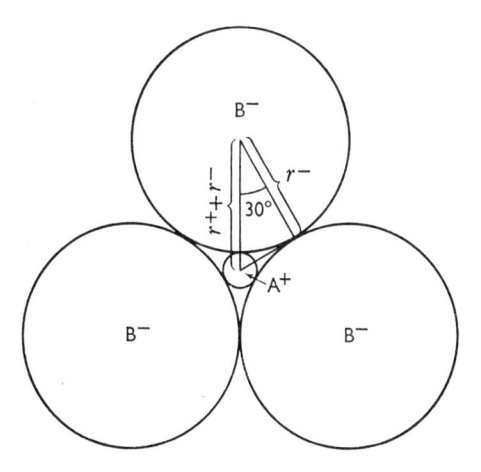

Fig. 18. The limiting r^+/r^- value for a triangular arrangement

with co-ordination numbers of 6 would be expected, and this, in fact, is what is found.

The way in which the ions build up into a sodium chloride crystal can be followed in stages. An ion pair, shown in Fig. 19, has a strong residual field and will attract a second ion-pair, just as two magnets would attract each other, as shown in Fig. 20. Four ion-pairs will arrange themselves as in Fig. 21, and a larger number will take up the arrangement shown in Fig. 22.

Other compounds, made up of simple ions, which have the same crystal structure as sodium chloride include the halides of lithium, sodium, potassium, rubidium and caesium (except the chloride, bromide and iodide of caesium), the oxides and sulphides of magnesium, calcium, strontium and barium, the chloride, bromide and iodide of silver, the monoxides of cadmium, iron(II), cobalt(II), nickel(II) and manganese(II), and the monosulphides of manganese (II) and lead(II).

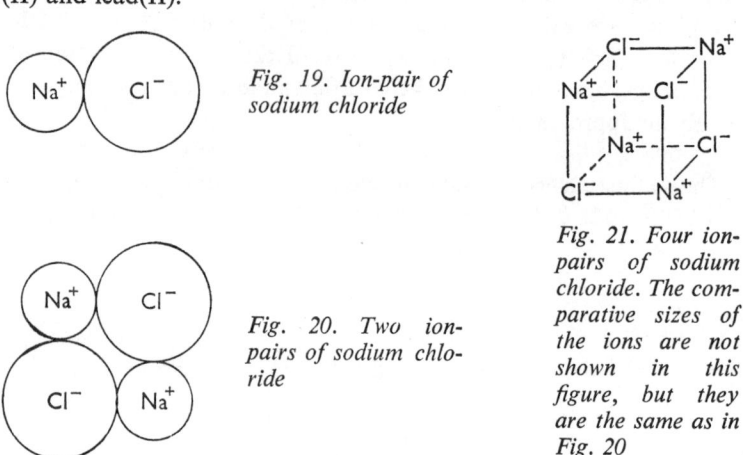

Fig. 19. Ion-pair of sodium chloride

Fig. 20. Two ion-pairs of sodium chloride

Fig. 21. Four ion-pairs of sodium chloride. The comparative sizes of the ions are not shown in this figure, but they are the same as in Fig. 20

One, or both, of the simple ions in a sodium chloride crystal may, also, be replaced by a charged group of atoms. Ammonium chloride, bromide and iodide, for instance, have the sodium chloride structure

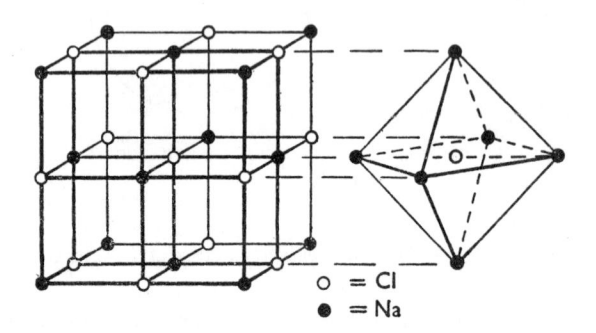

Fig. 22. Crystal structure of sodium chloride, showing (right) the octahedral arrangement of six sodium ions around one chloride ion

above their transition temperatures (page 54) whilst Ca^{2+} and CO_3^{2-} ions in calcite or Iceland Spar, or Ca^{2+} and $(C\equiv C)^{2-}$ ions in calcium carbide, are also arranged with the sodium chloride structure.

7 The caesium chloride crystal structure The radius ratio, Cs^+/Cl^-

for caesium chloride is 0·169/0·181 or 0·93 and the crystal structure is cubic with a co-ordination number of 8 (Fig. 23).

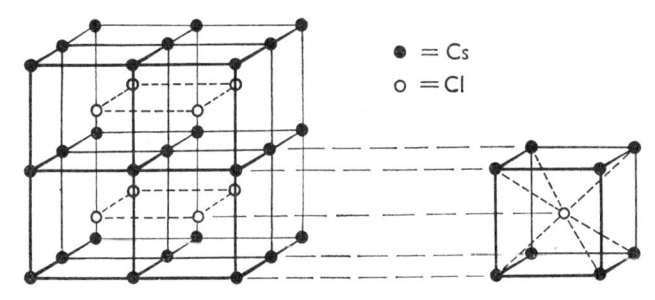

Fig. 23. Crystal structure of caesium chloride showing (right) *the cubical arrangement of eight caesium ions around one chloride ion*

Caesium bromide and iodide have the same structure, but caesium fluoride, with the smaller F^- ion, has a sodium chloride structure. Other substances which crystallise with a caesium chloride structure include caesium cyanide, thallium(I) cyanide, chloride, bromide and iodide, and the chloride, bromide and iodide of ammonium below their transition points (page 54).

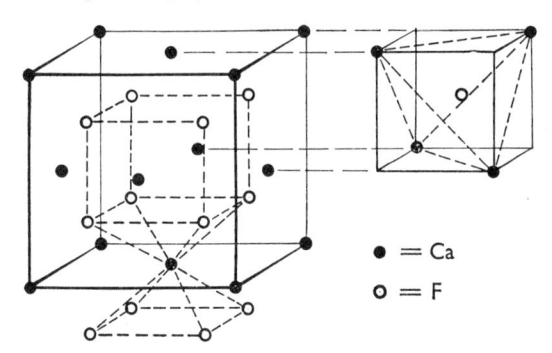

Fig. 24. Fluorite structure showing (right) *the tetrahedral arrangement of four calcium ions around one fluoride ion, and* (below) *the cubical arrangement of eight fluoride ions around one calcium ion*

8 Crystal structures of AB_2 ionic compounds The two commonest crystal structures for ionic compounds of this type are the fluorite, CaF_2, and the rutile, TiO_2, structures.

In the fluorite structure (Fig. 24) of CaF_2, each Ca^{2+} ion is surrounded by eight F^- ions at the corners of a cube, and each F^- ion is surrounded by four Ca^{2+} ions arranged tetrahedrally. The co-ordination numbers are 8 and 4.

In the rutile structure of TiO_2 (Fig. 25) each Ti^{4+} ion is surrounded by six O^{2-} ions arranged octahedrally, and each O^{2-} ion is surrounded by three Ti^{4+} ions arranged triangularly. The co-ordination numbers are 6 and 3.

A third closely related structure is the *anti-fluorite structure* which is the normal fluorite structure with the anions and cations interchanged.

Some common compounds crystallising in these forms are listed below:

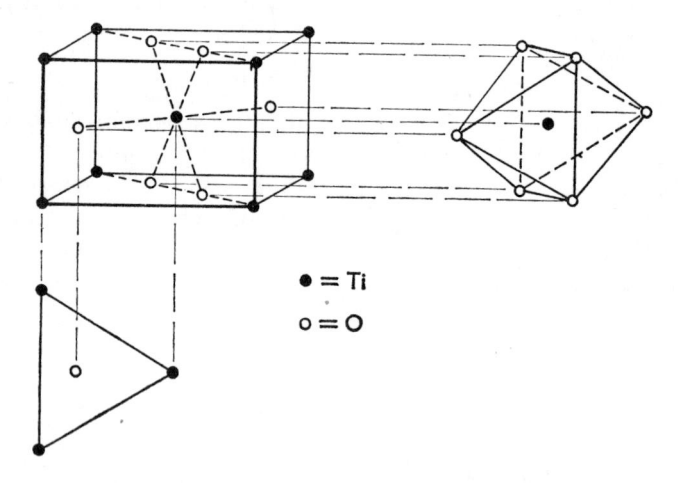

● = Ti
o = O

Fig. 25. Rutile structure showing (right) *the octahedral arrangement of six oxide ions around one titanium ion and* (below) *the triangular arrangement of three titanium ions around one oxide ion.*

Fluorite structure	Rutile structure	Anti-fluorite structure
Difluorides of calcium, strontium, barium, cadmium, and mercury(II)	Difluorides of nickel (II), cobalt(II), iron (II), manganese(II), magnesium and zinc Tin(IV) oxide Lead(IV) dioxide Manganese(IV) dioxide	Oxides and sulphides of lithium, sodium, potassium

ENERGY CHANGES IN THE FORMATION OF IONIC BONDS

9 Formation of an ionic bond from free atoms   In the formation of an ionic bond between two free atoms there are three factors to take into account so far as energy change is concerned:

(a) the ionisation energy in forming the cation;

(b) the electron affinity in forming the anion; and

(c) the electrostatic attraction between the two ions.

(a) and (b) can be measured experimentally (pages 44 and 48); (c) can be calculated if the two ions concerned are taken as charged spheres and if the internuclear distance between them is known (page 53).

If two spheres are imagined with charges $+e_1$ and $-e_2$, at a distance r apart, then the attractive force is given by Coulomb's law as $F = (e_1 \times e_2)/r^2$. If now the spheres are displaced through a very small distance dr, the force may be taken as remaining constant, and the work done in displacement is given by force × distance, i.e. $F.dr$.

In forming an ionic bond between two ions (represented as charged spheres) it is a matter of bringing the spheres to within a distance d of each other from a distance which by comparison is infinite. The energy released in doing this is, therefore, given by

$$\int_{\infty}^{d} F.dr, \quad \text{or} \quad \int_{\infty}^{d} \frac{e_1 \times e_2}{r^2}.dr$$

and this gives the result $(e_1 \times e_2)/d$.

In the particular case of the formation of an ionic bond between free atoms of sodium and chlorine, the overall energy change will be

$$\text{Ionisation energy of sodium } (I_{Na}) - \text{Electron affinity of chlorine } (E_{Cl}) - \frac{e^2}{r_{Na} + r_{Cl}}$$

where r_{Na} and r_{Cl} are the ionic radii of the sodium and chloride ions and e is the ionic charge. The process of bond formation will only take place if the energy change is negative, i.e. if the bonded ions are more stable than the free atoms, and this means that $e^2/(r_{Na}+r_{Cl})$ must be greater than $I_{Na} - E_{Cl}$. In the example quoted this is, in fact, the case and hence a bond is formed.

10 Crystal or lattice energy   In the preceding section the formation of an ionic bond from free atoms of sodium and chlorine was considered, but in the formation of a crystal of sodium chloride from solid sodium and gaseous chlorine other factors must be taken into account.

In the first place, the ion-pair resulting from one free atom of sodium and one free atom of chlorine has a strong residual field so that many ion-pairs build up into an ionic crystal. This involves further energy changes, and the total energy change when free ions form together into a crystal is known as the crystal or lattice energy. Alternatively this energy can be regarded as that required to break down a crystal structure into free ions.

Secondly, the formation of free sodium atoms from solid sodium, and of free chlorine atoms from gaseous chlorine, both demand the expenditure of some energy.

To calculate the heat of formation of sodium chloride, i.e. the heat of reaction of

$$Na(s) + \tfrac{1}{2}Cl_2(g) \rightarrow NaCl(c) + \underset{\text{(heat of reaction)}}{H}$$

it is necessary then to take all the following energy factors into account:

(a) The ionisation energy of sodium, I_{Na} (page 44).

(b) The electron affinity of chlorine, E_{Cl} (page 48).

(c) The crystal energy of sodium chloride, C_{NaCl}.

(d) The energy required to dissociate solid sodium into free atoms. This is equal to the heat of sublimation, S_{Na}.

(e) The dissociation energy of gaseous chlorine, D_{Cl}.

11 The Born-Haber cycle All these energy factors are interrelated in the Born-Haber cycle, which can be summarised as follows:

$$\begin{array}{ccc}
Na(s) + \tfrac{1}{2}Cl_2(g) & \xrightarrow{-H} & NaCl(c) \\
{\scriptstyle +S_{Na}}\downarrow \quad {\scriptstyle +D_{Cl}}\downarrow & & \uparrow {\scriptstyle -C_{NaCl}} \\
Na(g) + Cl(g) & \xrightarrow{I_{Na}+E_{Cl}} & Na^+(g) + Cl^-(g)
\end{array}$$

By Hess's law the energy change in the formation of crystalline sodium chloride must be the same however it is formed and it therefore follows that

$$H = -S_{Na} - D_{Cl} + I_{Na} + E_{Cl} + C_{NaCl}$$

By measuring H, S_{Na}, D_{Cl}, I_{Na}, and calculating C_{NaCl}, the equation enables values of E_{Cl} to be determined, and when this is done for a series of metallic halides, for instance, the constancy of the values obtained for E_{Cl} is remarkable.

Similarly, if E_{Cl} is measured experimentally along with S_{Na}, D_{Cl}, and I_{Na}, and C_{NaCl} is again calculated, it is possible to obtain theoretical values for H, and these again agree remarkably well with experimentally determined values.

Calculation of the crystal energy for some crystal structures is not easy, but experimental values can be obtained by employing the principle of the Born-Haber cycle if the other factors involved can be measured.

12 Repulsive forces between ion pairs The discussion of the energy changes in the formation of ionic bonds has been simplified particu-

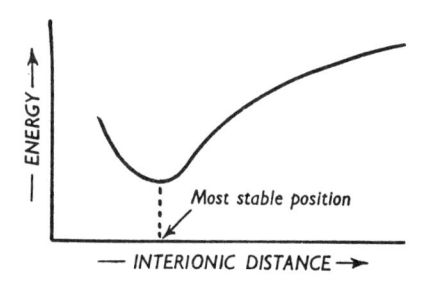

Fig. 26. The energy change as two ions are brought together

larly by regarding the ions as hard spheres and by neglecting any treatment of the repulsive forces which come into play as the electron shells of two ions begin to overlap. It is this repulsive force which opposes the attractive force between ions and results in the ions taking up an equilibrium position with a finite and measurable internuclear distance.

The energy change as two ions are brought closer together is indicated by the shape of the curve in Fig. 26. The shape is a result of the competing contributions of the attractive and repulsive forces to the energy of the system and shows that the ions actually arrange themselves at an internuclear distance which corresponds to the most stable position.

MAGNETIC MOMENTS OF IONS

13 Paramagnetism and diamagnetism Substances may be classified as paramagnetic or diamagnetic. Paramagnetic substances are drawn into a strong magnetic field, i.e. a rod of such a substance takes up a position parallel to the field (Fig. 27a). This is because a paramagnetic material is more permeable to magnetic lines of force than a vacuum. In other words, the permeability is greater than 1.

Diamagnetic substances tend to be drawn out of a magnetic field; rods of such substances set themselves at right angles to the field (Fig. 27b). They have a permeability less than 1.

Substances normally regarded as magnetic, e.g. steel, iron, cobalt, nickel and magnetic alloys, are paramagnetic, but the degree of magnetism possessed by such substances is much greater than that of any others; they are said to be ferromagnetic.

(a) Paramagnetic substance.   (b) Diamagnetic substance

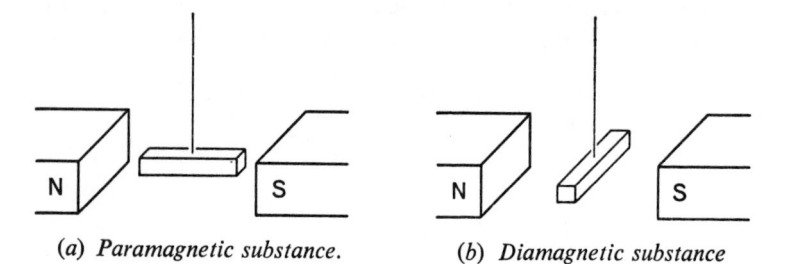

Fig. 27. *Effect of suspending paramagnetic and diamagnetic substances in a magnetic field. Paramagnetic substances take up a position parallel to the field; diamagnetic substances set themselves at right angles to the field*

14 Paramagnetism and unpaired electrons A substance or an ion in which all the electrons are paired (page 17) is diamagnetic, whilst substances or ions containing unpaired electrons are paramagnetic. This is because a spinning electron is equivalent to an electric current in a circular conductor and, as such, behaves as a magnet.

The magnetic moment of an electron is given by

$$\mu_B = \frac{he}{4\pi m}$$

where e is the charge on the electron in coulombs and m its mass in kg. When an orbit contains two paired electrons, the magnetic moment of one is compensated by the equal and opposite moment of the other. The unit of magnetic moment, known as the Bohr magneton, has a value given by the above expression, of $9 \cdot 274 \times 10^{-24}$ J T^{-1}.

The magnetic moment for an ion containing n unpaired electrons is equal to $\sqrt{(n(n+2))}$ magnetons, at least as an approximation and for many ions. This relationship can be used to find the number of unpaired electrons in an ion from measurements of its magnetic moment. The expected theoretical results are given below:

Number of unpaired electrons	0	1	2	3	4	5
Magnetic moment/Bohr magneton	0	1·73	2·83	3·87	4·90	5·92

The observed magnetic moments of the ions formed by the elements in the first transition series, and by the elements which precede

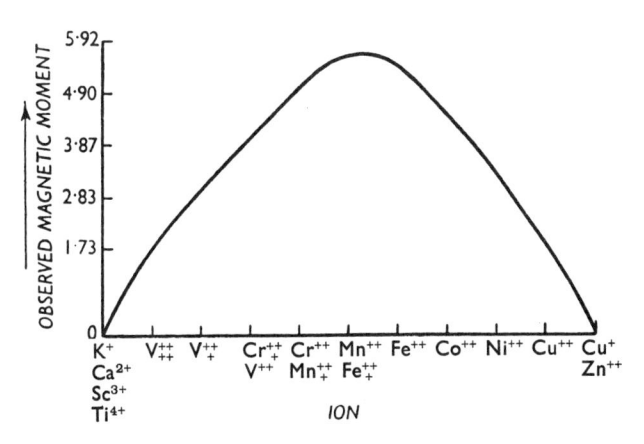

Fig. 28. *Observed magnetic moments of ions*

and follow this series, are shown graphically in Fig. 28. The results are in agreement with the electronic arrangements expected for these ions as shown in the following selected examples.

Ion	\multicolumn{6}{c}{Structure of ion}	Number of unpaired electrons
	1s	2s	2p	3s	3p	3d	
K⁺	2	2	2 2 2	2	2 2 2		0
V⁴⁺	2	2	2 2 2	2	2 2 2	1	1
V³⁺	2	2	2 2 2	2	2 2 2	1 1	2
V²⁺	2	2	2 2 2	2	2 2 2	1 1 1	3
Cr²⁺	2	2	2 2 2	2	2 2 2	1 1 1 1	4
Fe³⁺	2	2	2 2 2	2	2 2 2	1 1 1 1 1	5
Fe²⁺	2	2	2 2 2	2	2 2 2	2 1 1 1 1	4
Co²⁺	2	2	2 2 2	2	2 2 2	2 2 1 1 1	3
Ni²⁺	2	2	2 2 2	2	2 2 2	2 2 2 1 1	2
Cu²⁺	2	2	2 2 2	2	2 2 2	2 2 2 2 1	1
Cu⁺	2	2	2 2 2	2	2 2 2	2 2 2 2 2	0

This agreement between calculated and experimental results is evidence in favour of the validity of the rule of maximum multiplicity (page 27).

The measurement of magnetic moments is also of great value in elucidating the arrangements of electrons in complex ions, as explained on pages 166-8.

7 The Wave Nature of the Electron

1 Dual nature of light The word 'ray' is used somewhat indiscriminately to describe both a stream of particles and wave-like radiation, but the usage is, perhaps, valid when it is realised that particles may have a wave-like aspect whilst waves may have a particle-like aspect.

Newton (1642–1727) originally regarded light on a corpuscular theory, but Huyghens (1629–95) introduced the wave theory of light. The wave theory is essential in accounting for interference and diffraction phenomena; the corpuscular theory is necessary to explain, for instance, the photo-electric effect.

In this effect, metals give off electrons when illuminated with light of appropriate wavelengths. For the alkali metals, which lose electrons easily (page 45), light in the visible region can be used, and caesium, for example, is used in making photo-electric cells. For other metals, which lose electrons less easily, ultra-violet light is required to induce a photo-electric effect. For each metal there is a critical wavelength above which no photo-electrons are emitted.

The kinetic energy of the emitted electrons is inversely proportional to the wavelength of the incident radiation, but independent of the intensity of the radiation. The rate of emission of electrons is proportional to the intensity of radiation.

These experimental results were interpreted, by Einstein in 1905, as meaning that radiation could be regarded as made up of small 'packets' of energy, known as photons. When a metal was irradiated by such photons, some of the energy of the photons was used in ejecting electrons from the metal whilst the remainder was given up to the electrons.

The energy of a photon is dependent on the wavelength, or frequency, of the radiation concerned according to the basic equation of the quantum theory,

$$E = h\nu$$

which explains why the energy of the emitted electrons in the photo-electric effect is related to the frequency or wavelength of the radiation. A photon is a quantum of light energy. A photon of red light has energy of about $2 \cdot 5 \times 10^{-19}$ J; for ultra-violet light, the value is about 5×10^{-19} J.

2 The wave nature of an electron So far, the electron has been regarded as a tiny, negatively charged particle, but very important results come from a consideration of the wave nature of an electron. de Broglie first suggested, in 1924, that moving electrons had waves of definite wavelength associated with them, and this theoretical prediction was demonstrated experimentally when Davisson and Germer showed, in 1927, that a stream of electrons could be diffracted by crystals acting as simple diffraction gratings, just as light- or X-rays can. As it is only possible to account for diffraction in terms of waves, it is necessary to assume that a stream of electrons behaves as a wave-like radiation such as light- or X-rays. An electron microscope makes use of this fact.

The de Broglie relationship between moving electrons and the waves associated with them can be expressed mathematically as

$$\lambda = \frac{h}{mv}$$

where λ is the wavelength of the waves, h is Planck's constant, m is the mass of the electron, and v is the velocity of the electron. If this expression is written as

$$\text{Momentum} \times \lambda = h$$

it relates the particle-like aspect of an electron, i.e. momentum, to the wave-like aspect, i.e. wavelength. In this respect, Planck's constant is a proportionality constant.

The expression is true for all particles, but it is only with very small particles that the wave-like aspect is of any significance. The wavelength of a large particle will be so small that its wave-like properties will not be measureable or observable. An electron of energy 1 eV (page 14) has an associated wavelength of 1·2 nm; an α-particle from radium, a wavelength of $6·6 \times 10^{-6}$ nm; and a golf ball travelling at 30 m s^{-1}, a wavelength of $4·9 \times 10^{-25}$ nm.

It is not easy to obtain a pictorial idea of this new conception of an electron, but it is possible to treat the wave nature of an electron mathematically. This is done by a specialised technique known as wave mechanics or quantum mechanics.

3 Wave functions Schrödinger postulated, in 1927, that the wave pattern associated with an electron could be expressed in the form of a mathematical equation which involves a wave function (ψ), the total energy (W) and the potential energy (V) of the system, the mass of the electron (m), Planck's constant (h) and the co-ordinates of the system. The equation is written as

$$\frac{\delta^2\psi}{\delta x^2}+\frac{\delta^2\psi}{\delta y^2}+\frac{\delta^2\psi}{\delta z^2}+\frac{8\pi^2 m}{h^2}(W-V)\psi = 0$$

It cannot be proved from fundamental principles; it was Schrödinger's mathematical experience and intuition which enabled him to suggest the equation. It is now generally accepted as correct because the various solutions to the equation fit the known experimental facts. Indeed it was one of the early triumphs of this so-called wave equation that it accounted for the spectrum of hydrogen just as well as the Bohr theory of the atom, whilst it did better than the Bohr theory in accounting for the spectrum of helium.

The Schrödinger wave equation accounts for the facts because it can only be solved satisfactorily for certain, definite values of the total energy, W. These definite energy values, or eigenvalues, correspond to those of the stationary states in the Bohr model of the atom, and with whole-number values corresponding to quantum numbers. For each definite energy value for which the equation can be solved there is an associated relationship giving the values of ψ, called the eigenfunctions, which are associated with the particular energy value.

So far as atomic structure is concerned, the idea of a tiny, negatively charged particle existing in particular orbits around the nucleus of an atom must be replaced. Instead, an electron must be thought of as existing in a much more diffuse region around the nucleus. This region is called an atomic orbital, and it is the shape of atomic orbitals which is of major importance. Three different types of orbital are mainly concerned; they are known as the s-, p- and d-orbitals. f-orbitals also exist but are, as yet, of minor importance.

Wave functions are important because they can be used to indicate the shape, and other properties, of s-, p- and d-atomic orbitals.

4 The significance of ψ and ψ^2 The nature of the Schrödinger equation is such that the wave function, ψ, may be regarded as an amplitude function. This corresponds, on a three-dimensional scale, to the function which expresses the amplitude of vibration of a plucked string.

As it is the square of the amplitude of the vibrating string which measures the intensity of the wave involved, so ψ^2 measures the probability of an electron existing at a point or within a small volume of space. A zero value of ψ^2 at a fixed point, means, therefore, that the chance of finding an electron at that point is zero. A high value of ψ^2 at a point means that there is a high chance of finding an electron at that point.

The value of the wave function, ψ, for an electron in an atom is dependent, in general, both on the radial distance, r, of the electron from the nucleus of the atom and on its angular direction away from the nucleus.

The nucleus, itself, however, cannot provide any sense of direction until a set of arbitrarily chosen axes are superimposed. If Cartesian axes are chosen, as in Fig. 29a, then a point in space, P, around the nucleus, N, can be defined in terms of x-, y- and z-co-ordinates. Such axes do not, however, have any absolute directional significance until an external magnetic field is applied. The axes then have a definite direction in relation to the direction of the magnetic field. Alternatively, polar co-ordinates, as in Fig. 29b, can be used.

(a) Cartesian coordinates   (b) Polar coordinates

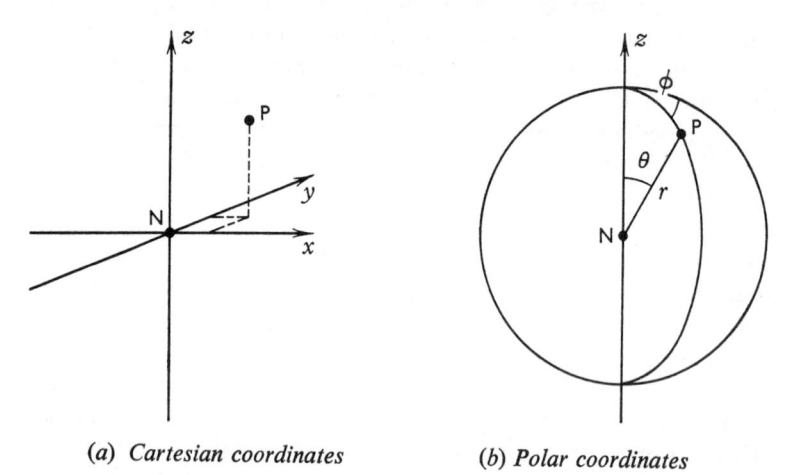

Fig. 29. *The position of a point* P *relative to point* N *can be expressed in terms of* (a) *Cartesian coordinates, x, y and z, or* (b) *polar coordinates, r, θ and φ*

The mathematical relationship between ψ and r and the angular direction can be established accurately for a single electron in a hydrogen-like atom. For more complicated atoms, containing more than one electron, the corresponding relationships can only be established approximately because of the difficulty involved in solving the mathematical equations.

The differences between s-, p- and d-orbitals depend on the different ways in which ψ and/or ψ^2 vary with r and with the angular direction, as explained in the following sections.

5 s-orbitals There are three useful ways of considering s-orbitals.

(a) *Change of ψ with r.* The ψ value for an electron in a 1s-orbital is dependent on r, but independent of the angular direction away from the nucleus. The value of ψ falls rapidly as r increases, as shown in Fig. 30a. The state of affairs can also be represented by drawing

(a) *Variation of ψ with r*

(b) *Spherical contours of constant ψ value around the nucleus*

(c) *Boundary surface within which ψ has a greater value than an arbitrarily chosen value*

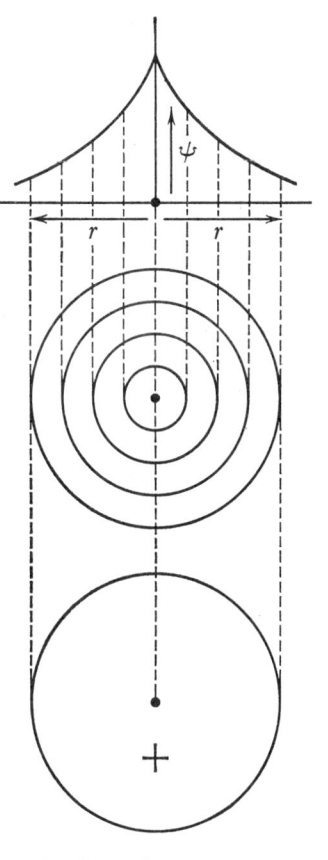

Fig. 30. Three ways of representing a 1s orbital graphically, all based on values of ψ

spherical contour lines around the nucleus of the atom, each contour showing a constant ψ value (Fig. 30b). Alternatively, a spherical boundary surface can be drawn within which ψ has a value greater than any arbitrarily selected value (Fig. 30c). Each of these methods provides a graphical representation of a 1s-orbital. The boundary surface method can also be used to show the sign of the wave function, by placing a + in the circle.

For a 2s-orbital, ψ becomes zero and then negative, as r increases, as shown in Fig. 31a. The corresponding boundary surfaces, shown in Fig. 31b, represent a 2s-orbital, and also show the positive and negative values of ψ. Whenever ψ has a zero value it means that there is no chance of finding an electron. The dotted circle in Fig. 31b shows, therefore, a spherical surface in which there is no chance of finding an electron. Such a surface is known as a *nodal sphere or surface*.

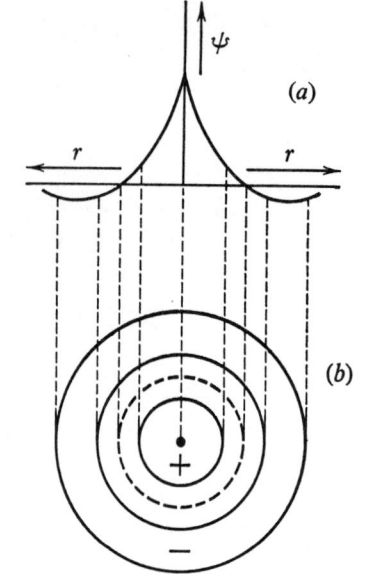

Fig. 31. Representation of a 2s orbital, based on ψ values, by (a) graph showing change of ψ with r, and (b) boundary surfaces. In (b), the dotted circle shows the nodal sphere

All s-orbitals have a spherical shape like the 1s- and 2s-orbitals; they are said to be *spherically symmetrical*. They differ by the number of nodal surfaces, the *ns*-orbital having $(n-1)$ of them.

(b) *Change of ψ^2 with r.* The value of ψ drops rapidly as r increases, for a 1s-orbital. The value of ψ^2 must also drop as r increases, as shown in Fig. 32b.

Because the value of ψ^2 at any point is related to the chance of finding an electron at that point, it is possible to build up a simple pictorial interpretation of the graph shown in Fig. 32a in terms of probability.

If a single electron in a 1s-orbital could be photographed on the same film in each of the many positions it might be said to occupy over a period of time, then the photograph would show the probability distribution of the electron. There would be the greatest 'electron density' in those regions where the electron had the highest probability of existing, and vice versa. The overall picture

(a) *Charge cloud round nucleus, showing decrease in 'electron density' as r increases*

(b) *Decrease in ψ^2 as r increases*

(c) *Boundary surface*

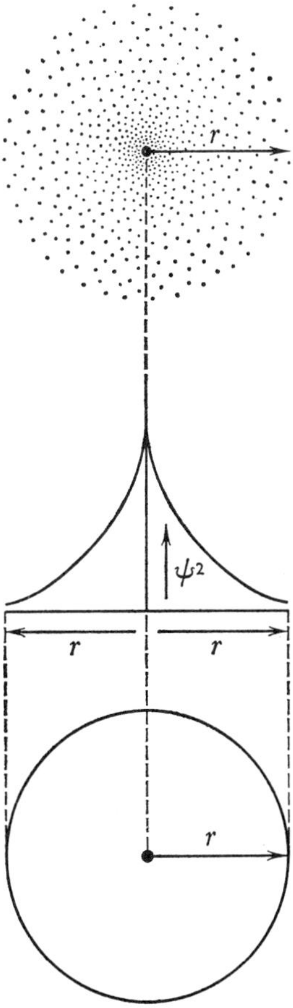

Fig. 32. *Probability distribution for a 1s atomic orbital, based on ψ^2 values*

would be one of a charge cloud around the nucleus with the 'electron density' falling off quite rapidly in moving away from the nucleus in any direction. This charge cloud is represented, for a 1s-orbital, in Fig. 32a.

There is a chance of finding an electron in a 1s-orbital a long way from the nucleus, but the chance is very, very slight because ψ^2 falls off rapidly as r increases as shown in Fig. 32b. It is, therefore, possible to draw a sphere, known as a boundary surface, around the nucleus to represent the probability distribution of a 1s-orbital (Fig. 32c). The radius can be chosen arbitrarily to give a 90 or 95 or n per cent probability that the electron will be found within the sphere. The

(a) *Charge cloud round nucleus showing change in 'electron density' as r changes. The nodal sphere is dotted*

(b) *Change in ψ^2 as r increases*

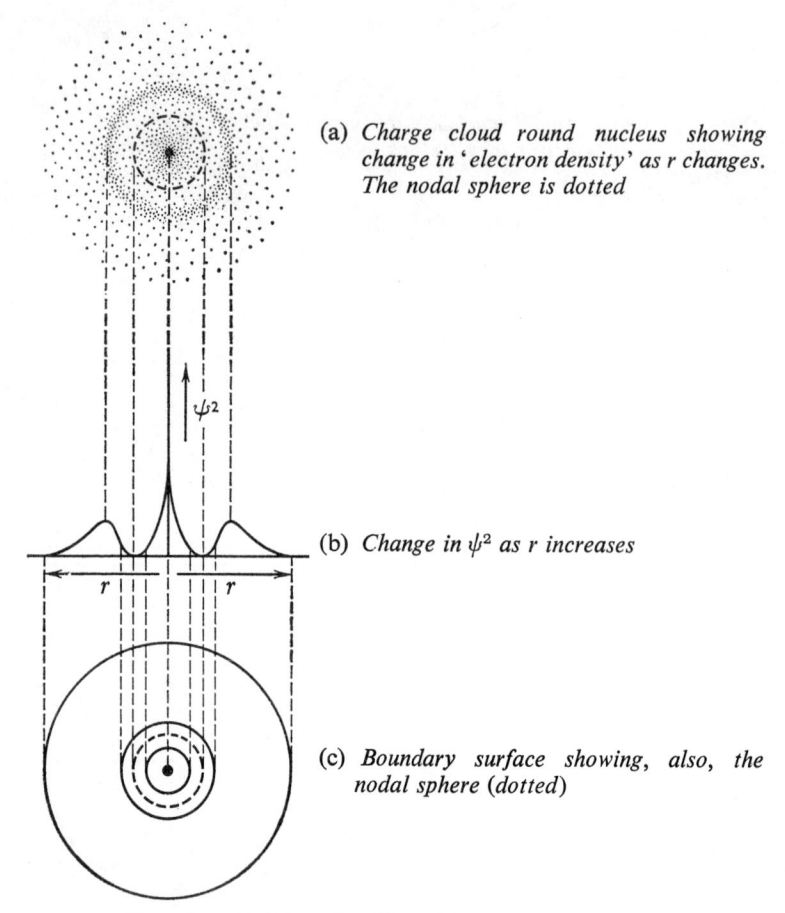

(c) *Boundary surface showing, also, the nodal sphere (dotted)*

Fig. 33. *Probability distribution for a 2s atomic orbital*

precise value of r which is chosen is not very significant for non-mathematical purposes. What does matter is that the probability distribution for s-orbitals is spherically symmetrical.

The change in ψ^2 with r, the charge cloud picture and the boundary surfaces corresponding to a 2s-orbital are shown in Fig. 33. Whereas ψ can have positive, zero or negative values (page 112), the value of ψ^2 is always positive or zero, which it must be if it is to represent a probability. Probability distribution diagrams do not, therefore, show + or − signs.

It is important to realise the distinction between an orbital and the corresponding probability distribution. The shape of an orbital is based on the variation of ψ with r, and ψ is a mathematical function which cannot be given a full pictorial significance. The shape of a probability distribution is based on the variation of ψ^2 with r. This can be pictured in terms of a charge cloud, because ψ^2 represents a probability. For s-orbitals the shape of orbital and probability distribution are both spherical, but this is not so for p- and d-orbitals (pages 116–18).

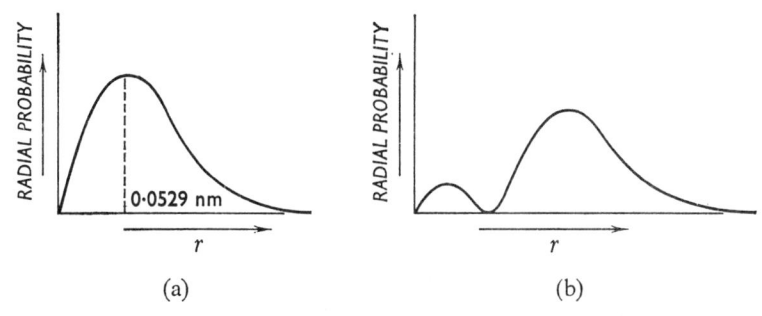

Fig. 34. The change of radial probability with r for (a) *a 1s orbital,* (b) *a 2s orbital. In* (a) *the maximum radial probability occurs when r = 0·0529 nm, which is the radius of the innermost stationary state in the Bohr model*

(c) *Radial probability distribution.* The value of ψ^2 *at a point* gives the probability of finding the electron at that point or within a small region of volume around the point. As will be seen from Fig. 32b, this probability is greatest nearest the nucleus.

The probability of finding the electron within a spherical shell, of thickness dr, at a distance r from the nucleus is rather different, and is known as the *radial* probability. This is because the actual volume of the shell increases as r increases, whilst the probability of finding the electron in a *fixed* volume of space decreases as r increases (Fig. 32b).

The two factors taken together mean that the radial probability for a 1s-orbital changes with r as shown in Fig. 34a. It will be seen that the greatest radial probability occurs when r equals 0·0529 nm, and this is the value of the radius of the innermost stationary state of a hydrogen atom as calculated by Bohr. Consideration of probability distribution together with radial probability therefore brings about a correlation between the charge cloud picture and the earlier stationary states of the Bohr model of the atom.

The variation of radial probability with r for a 2s-orbital is shown in Fig. 34b.

6 p-orbitals The ψ value for an s-orbital does not depend on the angular direction away from the nucleus; that is why s-orbitals are spherically symmetrical. The ψ value for an s-orbital does, however, depend on r as illustrated in Fig. 30.

The solution of the wave equation shows that the ψ value for p-orbitals depends both on r *and* on the angular direction away from the nucleus. This means that a full diagrammatic representation of a p-orbital is impossible; four dimensions would be needed to show the variation of ψ with r, θ and ϕ (Fig. 29).

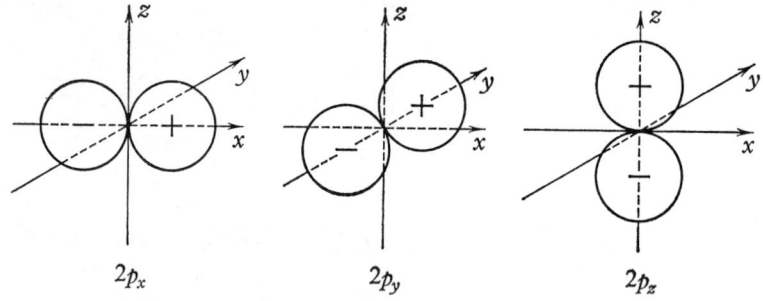

Fig. 35. Boundary surfaces for $2p_x$, $2p_y$, and $2p_z$ orbitals

It is, therefore, necessary to consider the variation of ψ with r separately from the variation of ψ with θ and ϕ. The radial probability distribution for a 2p electron is not unlike that for a 1s electron (Fig. 34a): the probability is zero at the nucleus and rises to a maximum before falling asymptotically. For a 3p electron the radial probability distribution is similar to that of a 2s electron (Fig. 34b), i.e. there are two maxima at different distances from the nucleus. The maxima for *ns* electrons are nearer to the nucleus than those for *np* electrons.

The radial probability distribution indicates the probability of finding an electron in a thin spherical shell at any distance, r, from the nucleus. It does not, however, indicate that, for p-electrons, this probability is very much greater in some directions than in others, because p-orbitals are not spherically symmetrical.

The directional nature, or angular dependence, of p-orbitals can be shown by plotting ψ on a polar diagram. The result is shown in Fig. 35 for the three degenerate p-orbitals, $2p_x$, $2p_y$ and $2p_z$. The directional nature of the orbitals along and around the x-, y- and z-axes is clear: the $+$ and $-$ values of ψ are also shown.

The corresponding diagram using ψ^2, instead of ψ, are shown in Fig. 36. These diagrams indicate the probability distribution so far

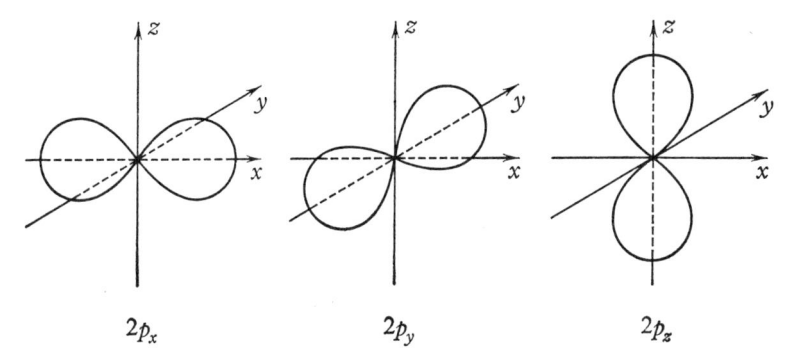

Fig. 36. Probability distribution for $2p_x$, $2p_y$, and $2p_z$ orbitals.

as the angular dependence is concerned. It will be seen that a p_x-orbital has a nodal plane (p. 72) in the yz-plane; a p_y-orbital has a nodal plane in the xz-plane; and a p_z-orbital has a nodal plane in the xy-plane.

Neither the diagram in Fig. 35 nor that in Fig. 36 indicates r, i.e. the distance away from the nucleus. To get a complete picture of a

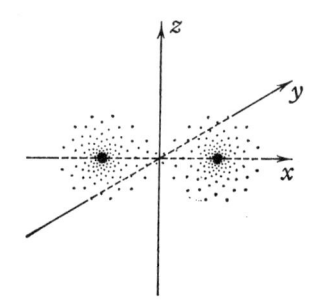

Fig. 37. Charge cloud for a $2p_x$ orbital

p-orbital involves a combination of the angular dependence and the radial dependence. As it is, usually, the directional nature of p-orbitals that is of greatest importance the diagrams in Figs. 35 and 36 are commonly used to represent p-orbitals. It must, however, be realised that they do this incompletely, as do the diagrams such as that in Fig. 37 which attempt to portray charge-cloud pictures of p-orbitals.

7 d- and f-orbitals For each principal quantum number there are s-orbitals, which are spherically symmetrical. For each principal quantum number greater than 1 there are also three degenerate (page 18) p-orbitals directed mutually at right angles to each other along the x-, y- and z-axis but similar in shape.

For each principal quantum number greater than 2 there are also five degenerate d-orbitals, which are not, however, all of the same shape. In one set, of three, the lobes of the orbitals lie along lines making angles of 45° to the x-, y- and z-axes, i.e. the lobes fall between the axes (Fig. 38). These three orbitals are labelled d_{xy}, d_{xz} and d_{yz} according to the plane in which the orbital lies. This set of three orbitals is known collectively as the t_{2g} or d_ϵ-orbitals.

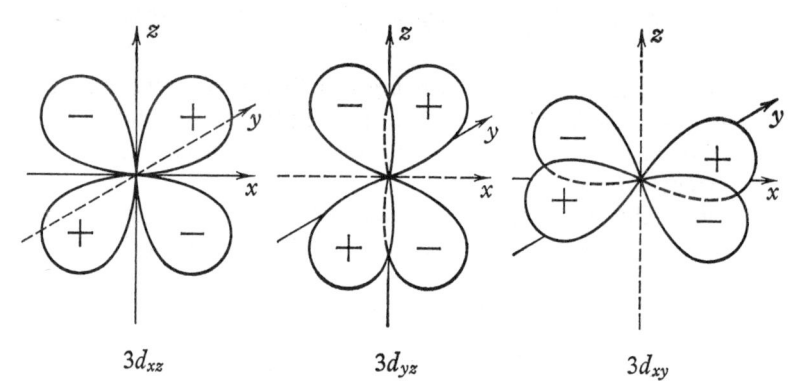

Fig. 38. The t_{2g} or d_ϵ set of 3d orbitals

In the second set of two orbitals (Fig. 39) the lobes lie either along the x- and y-axes or along the z-axis. The first of these orbitals is referred to as the $d_{x^2-y^2}$-orbital; the second as the d_{z^2}-orbital. The two orbitals together are known as the e_g- or d_γ-orbitals.

As will be seen (Chapter 16), the d-orbitals play an important part in the bonding in complexes of the transition metals.

For principal quantum numbers greater than 3 there are also seven degenerate f-orbitals but they do not play a very large part in present-day theories of chemical bonding.

8 The uncertainty principle One of the fundamental differences between an electron and a large particle is that the wave-like aspect of the electron is of much greater significance than the wave-like aspect of the larger particle (page 68).

There is, too, a further difference. Both the position and the velocity of a large particle, e.g. a planet, at any one time can be measured with fair accuracy, but this cannot be done for anything so small as an electron. There is simply no way of measuring the velocity of an electron exactly and of locating it exactly at the same time. This is because any method of measurement affects the electron being measured. An electron might be detected, for example, by using very short wavelength X-rays, or γ-rays, if the electron would cause scattering of the rays, just as a speck of dust can be detected in a beam of sunshine. But the speed and direction of movement of the electron would be affected by the rays, whereas light has little or no

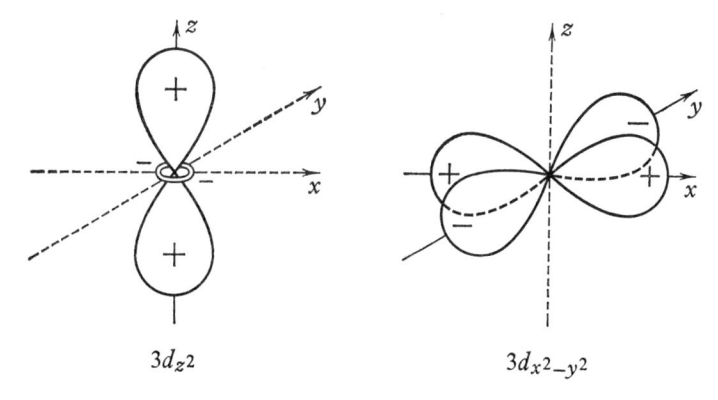

Fig. 39. The e_g or d_γ set of two 3d orbitals

effect on a dust particle. Electrical or mechanical methods would also affect the speed and direction of movement of an electron.

The information which can be obtained about an individual electron is, therefore, far from precise, and this is one example of the application of the Uncertainty Principle put forward by Heisenberg in 1927. This states that *the more accurately the position of a particle is defined the less accurately is its velocity known, and the more accurately the velocity is defined the less accurately is its position known.*

de Broglie's relationship (page 68) indicates why this is so. A long wavelength can be measured with greater fractional accuracy than a short one. If, therefore, a particle has a small momentum, and a correspondingly large wavelength, the wavelength can be measured with some accuracy but only at the expense of a relatively inaccurate determination of the small momentum. Alternatively, if the momentum is large and can be measured with some accuracy, the wavelength will be small and will not be known with any accuracy.

Because it is impossible to know the position and the velocity of any one electron, the best that can be done is to try to determine the probable position and the probable velocity. Fortunately, wave mechanics enables the various probabilities to be calculated, but, in so doing, the simple physical picture of a single particle-like electron moving with known velocity in a definite and precise orbit is lost. Just as the characteristics of any one individual are lost when an insurance company works out probabilities of life or death for, say, 1 000 000 people aged 25 years.

8 Molecular Orbitals

1 Introduction It is clear that ions are held together in an ionic compound by electrostatic attraction between positively and negatively charged ions, but it is less easy to understand why a covalent bond should hold two atoms together. The best understanding comes from a consideration of molecular orbitals.

Any molecule is supposed to have orbitals associated with it in much the same way as a single atom has. Just as each electron in a single atom can be represented by a wave function, ψ, denoting a particular atomic orbital, so each electron in a molecule can be represented by similar wave functions denoting particular molecular orbitals.

The various molecular orbitals associated with a molecule have different energy values and may have different shapes. They serve the same purpose as atomic orbitals in giving, through ψ^2 values, the relative probable electron densities in space. The Pauli principle (page 16) applies to molecular orbitals as to atomic orbitals, so that no single molecular orbital can contain two precisely similar electrons. This means that any particular molecular orbital can only contain two electrons and that these electrons must have opposed spins.

The electrons associated with a molecule, excepting those in the inner shells of the atoms concerned, are supposed to enter molecular orbitals and these orbitals fill up according to energy and spin considerations, just as atomic orbitals do (page 25).

The nomenclature, s, p and d used for atomic orbitals is replaced by that of σ, π and δ for molecular orbitals. δ-orbitals (page 87) are not, at this level of treatment, very important, but it is very necessary to have a clear idea as to the meaning and nature of σ- and π-molecular orbitals.

2 LCAO approximation The essential difference between an atomic orbital and a molecular orbital is that an electron in the former is influenced by only one positive nucleus whereas an electron in the latter is influenced by two, or more, nuclei. It is, nevertheless, reasonable to assume that molecular and atomic orbitals will have much in common; they both seek to describe the nature of an electron within a nuclear field.

The most convenient way of working out the wave functions for molecular orbitals is to adopt a method known as the linear combination of atomic orbitals (LCAO) approximation. Like many of the methods of wave mechanics it is only an approximate method simply because greater accuracy involves equations which cannot be formulated and/or solved.

The linear combination of two atomic orbital wave functions can be brought about either by adding or by subtracting the two wave functions. This can be expressed, in a simplified form, by the equation

$$\psi = \psi_A \pm \lambda . \psi_B$$

where ψ is the molecular orbital wave function obtained by the combination of atomic wave functions, ψ_A and ψ_B. λ gives a measure

Fig. 40. The boundary surface of a σ1s molecular orbital formed from two 1s atomic orbitals

of the ionic character of the bond between A and B (page 122) and its value is dependent on the nature of A and B. For a diatomic molecule of an element, A_2, ψ_A and ψ_B are equal, and the value of λ is 1.

3 Combination of s orbitals If two similar 1s-atomic orbitals be combined by addition, as, for example, in a hydrogen, H_2, molecule, the boundary surface of the resulting molecular orbital will be as shown in Fig. 40. Combination of the two corresponding atomic

Fig. 41. The ψ–r curve for the σ1s molecular orbital formed by combination of the two ψ–r curves for 1s atomic orbitals

ψ–r curves gives the approximate ψ–r curve for the molecular orbital (Fig. 41), and the orbitals are represented in terms of charge clouds in Fig. 42.

Figure 42 shows that there is an accumulation of negative charge between the nuclei. It is this which holds the nuclei together, at an equilibrium inter-nuclear distance (page 88), and which constitutes the covalent bond previously denoted by H:H or H–H. The molecular orbital is said to be a *bonding orbital*. It is called a σ1s-orbital, σ because it is symmetrical about the molecular axis, and 1s because it is formed by combination of 1s-atomic orbitals. Alternatively, the σ notation can be taken to mean that the molecular orbital has no nodal plane containing the molecular axis (page 87). The bond formed between two atoms by a σ-molecular orbital is called a *σ-bond*.

Fig. 42. The two 1s atomic orbitals and the resulting σ1s molecular orbital represented in terms of charge clouds

Combination of two similar 1s-atomic orbitals by subtraction is most easily brought about by adding the two 1s-atomic orbitals after changing the sign of one of them. The resulting $\psi - r$ curves and boundary surfaces are shown in Fig. 43. It will be seen that negative charge is withdrawn from the region between the nuclei and this

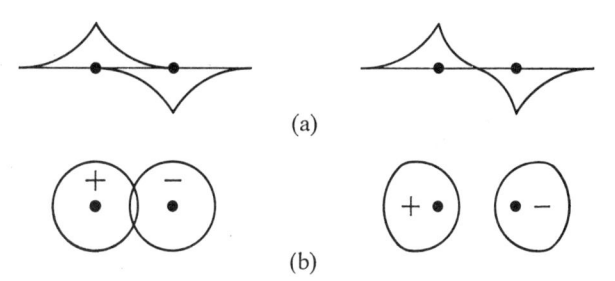

Fig. 43. Formation of anti-bonding σ*1s molecular orbital by combination of two 1s atomic orbitals by subtraction. (a) Showing the $\psi-r$ curves. (b) Showing the boundary surfaces

increases the repulsive forces between the nuclei. That is why this molecular orbital is an *anti-bonding orbital*; it is labelled σ*1s.

A combination of two atomic orbitals must give two molecular orbitals to accommodate the available electrons. Each atomic orbital and each molecular orbital can contain two electrons with opposed spin. One of the molecular orbitals is a bonding orbital: the other is anti-bonding (page 88).

4 Combination of p orbitals   p orbitals can overlap, and be combined either end-on, to give σ-molecular orbitals, or broadside on, to give π-molecular orbitals.

(*a*) *To form σ-molecular orbitals.* Two $2p_x$-atomic orbitals overlapping end-on can form a bonding molecular orbital as shown in Fig. 44.†

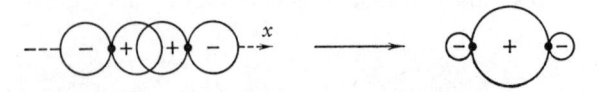

Fig. 44. The $\sigma 2p_x$ bonding orbital formed from two $2p_x$ orbitals

The build-up of negative charge between the nuclei, similar to that when *s*-orbitals form a bonding orbital, constitutes a σ-bond. The molecular orbital is labelled $\sigma 2p_x$.

The corresponding anti-bonding orbital, called $\sigma^* 2p_x$, is shown in Fig. 45.

Fig. 45. The $\sigma^* 2p_x$ anti-bonding orbital formed from two $2p_x$ orbitals

(*b*) *To form π-molecular orbitals.* *p*-orbitals can also combine broadside-on. If two $2p_z$-orbitals be considered, they can form a bonding orbital, shown in Fig. 46, and an anti-bonding orbital, shown in Fig. 47.

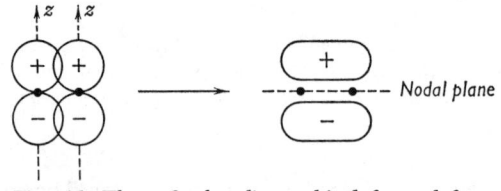

Fig. 46. The $\pi_z 2p$ bonding orbital formed from $2p_z$ orbitals

The bonding orbital is known as the $\pi_z 2p$ molecular orbital the anti-bonding orbital is labelled $\pi_z^* 2p$. The π_z nomenclature signifies

† It is convenient to place the *x*-axis along the line joining the nuclei of a diatomic molecule.

that the orbitals are not symmetrical about the molecular axis. Alternatively, it can be interpreted as meaning that the orbital has one nodal plane containing the molecular axis (page 87).

Electrons in a bonding π-orbital form what is known as a *π-bond*. The accumulation of negative charge holding two atoms together is *alongside* the molecular axis. In a σ-bond, the negative charge is between the two atomic nuclei.

5 Limitations to combinations of atomic orbitals The molecular orbitals formed by combining s-atomic orbitals and p-atomic orbitals have been described in the preceding section, but atomic orbitals may only be combined within certain limitations, outlined below:

(*a*) The energies of the atomic orbitals must be similar in magnitude. The application of this principle means that 1s- and 2s-, or s- and p-orbitals do not combine to form molecular orbitals in homonuclear diatomic molecules, i.e. A_2. The energy differences between 1s- and

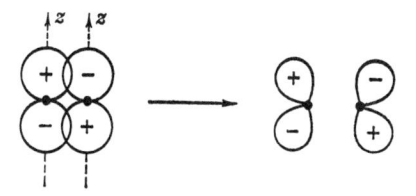

Fig. 47. The π_z^*2p anti-bonding orbital formed from two $2p_z$ orbitals

2s-, or 2s- and 2p-atomic orbitals are too great. In heteronuclear molecules, such as AB, this might not be so.

(*b*) The atomic orbitals must overlap to a considerable extent if they are going to combine to form a molecular orbital. The greater the overlap of the atomic orbitals concerned the greater the build-up of charge between the nuclei. This general idea is developed into the principle of maximum overlap on page 104.

(*c*) The atomic orbitals must have the same symmetry about the molecular axis. This means that some atomic orbitals of comparable energy which do overlap can still not be combined to give molecular orbitals. A $2p_z$-atomic orbital, for example, will not combine with an s-atomic orbital.

Consider an s-atomic orbital for an atom A, and a $2p_z$-atomic orbital for an atom B. If the molecular axis is in the x-direction, the expected overlap would be as shown in Fig. 48. But no molecular orbital results because the symmetry of the s-atomic orbital of atom A about the molecular axis AB is not the same as the symmetry of the $2p_z$-atomic orbital of atom B. Alternatively, it can be said that the $++$ overlap is neutralised by the equal $\pm$ overlap.

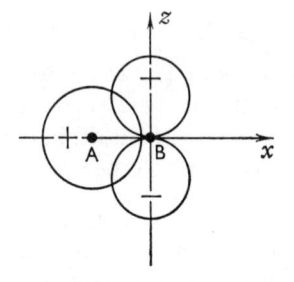

Fig. 48. A $2p_z$ orbital will not combine with an s orbital because they have not got the same symmetry about A–B

Fig. 49. A $2p_x$ orbital will combine with an s orbital

A similar argument applies to s- and $2p_y$-orbitals. A $2p_x$-orbital can, however, combine with an s-orbital as shown in Fig. 49 to form a σ-molecular orbital.

It is also evident that a p_z-orbital on one atom cannot be combined with a p_x- or a p_y-orbital on another.

6 Permitted combinations of atomic orbitals The atomic orbitals which can be combined together, from a symmetry point of view, are tabulated below:

First orbital	Second orbital	Type of molecular orbital formed
s	s p_x $d_{x^2-y^2}$ d_{z^2}	σ
p_x	s p_x $d_{x^2-y^2}$ d_{z^2}	σ
p_y	p_y d_{xy}	π
p_z	p_z d_{xz}	π
$d_{x^2-y^2}$	s p_x $d_{x^2-y^2}$ d_{z^2}	σ
d_{z^2}	s p_x $d_{x^2-z^2}$ d_{z^2}	σ
d_{xy}	p_y d_{xy}	π
d_{xz}	p_z d_{xz}	π
d_{yz}	d_{yz}	δ

It will be seen that there are three types of molecular orbital that can be formed by combination of these atomic orbitals. The main difference is that a σ-orbital has no nodal plane (page 83), a π-orbital has one nodal plane (page 84), and a δ-orbital has two nodal planes. The difference is very noticeable in the end-on views of the three types of orbital given in Fig. 50.

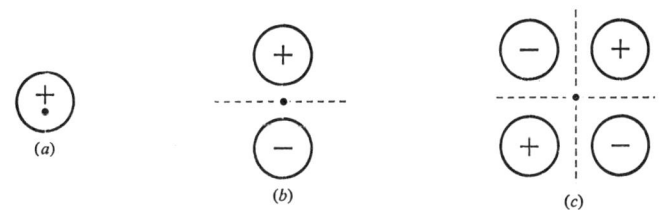

Fig. 50. Diagrammatic end-on views of (a) a σ-orbital, (b) a π-orbital, and (c) a δ-orbital. Nodal planes are shown by dotted lines

The formation of δ-orbitals, shown schematically in Fig. 51, does not play any large part in a discussion of chemical phenomenon, but the combination of d-orbitals to form σ- and π-molecular orbitals is of vital importance in the chemistry of the transition metal complexes (page 198). The chemistry of non-transition metals is, in the main, concerned with s- and p-orbital combinations.

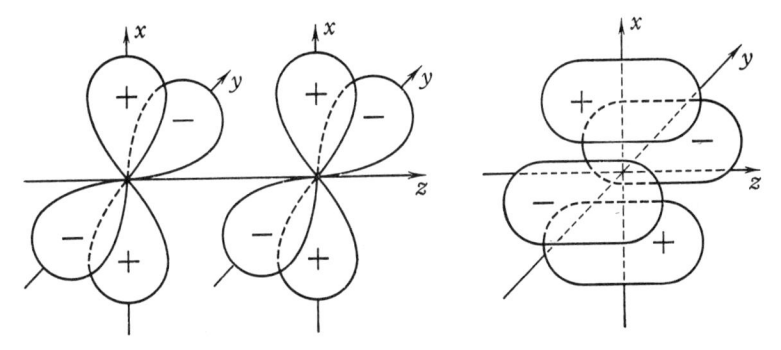

Fig. 51. Combination of two $d_{x^2-y^2}$ atomic orbitals to form a δ molecular orbital

7 Energy level of molecular orbitals The energy level of a bonding molecular orbital is always lower than that of the atomic orbitals from which it is derived; that is what makes the molecular orbital bonding. Conversely, the energy level of an anti-bonding molecular orbital is always higher than that of its corresponding atomic orbitals. This is illustrated in Fig. 52.

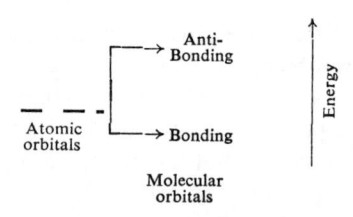

Fig. 52. The energy level of bonding and anti-bonding molecular orbitals in relation to that of the atomic orbitals from which they are formed

Alternatively, the total energy of a diatomic molecule, *AB*, can be plotted against the internuclear distance between *A* and *B*. In this way, the change in energy as atoms *A* and *B* are brought together can be seen (Fig. 53).

Line 2 shows a pronounced minimum and this represents the formation of a bonding molecular orbital. The internuclear distance

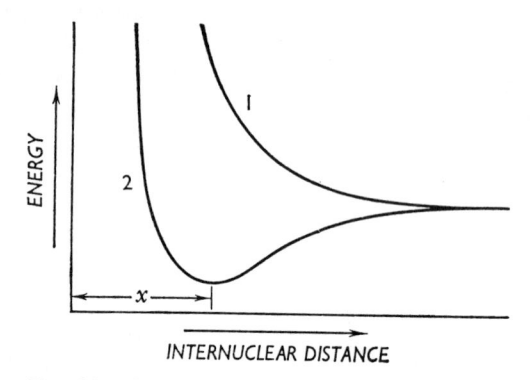

Fig. 53. The energy changes as two atoms are brought together to form a diatomic molecule

between the two bonded atoms is given by x, and the minimum in the curve arises from a combination of the attractive forces between positive nuclei and negative electrons taken together with the repulsive forces between nuclei and between electrons. At internuclear distances less than x, the repulsion of one nucleus by the other begins to predominate. At internuclear distances greater than x, the attractive force of the nucleus of one atom on the negative electronic charge of the other decreases.

Line 1 shows no minimum. It represents the formation of an anti-bonding molecular orbital. Repulsive forces both between nuclei and between electrons predominate and increase steadily as the internuclear distance decreases. So far as simple molecules of elements are concerned, the only molecular orbitals which need be considered are as follows:

| Bonding orbitals | $\sigma 1s$ | $\sigma 2s$ | $\pi_y 2p$ | $\pi_z 2p$ | $\sigma 2p$ |
| Anti-bonding orbitals | $\sigma^* 1s$ | $\sigma^* 2s$ | $\pi_y^* 2p$ | $\pi_z^* 2p$ | $\sigma^* 2p$ |

and the order of increasing energy for these orbitals, determined by spectroscopic measurements, is shown in Fig. 54.*

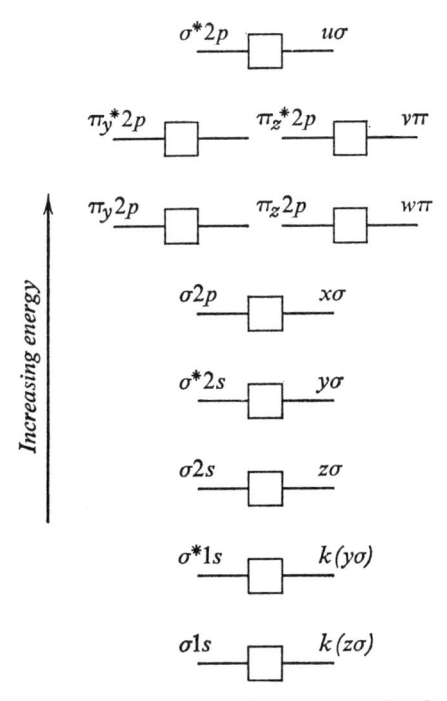

Fig. 54. Energy order of molecular orbitals for simple molecules of elements. The diagram is not to scale. The Mulliken system of nomenclature is shown on the right. Each orbital can hold two electrons.

This sequence enables the electronic structures of simple molecules to be worked out (page 91), for the available molecular orbitals 'fill up' in a molecule just as the available atomic orbitals 'fill up' in an atom. The aufbau principle (page 26) repeats itself.

* See footnote on p. 93.

8 Alternative system of nomenclature An alternative system of nomenclature, due to Mulliken, designates molecular orbitals according to their order in the energy sequence and to their σ- or π-type. The sequence

$$\sigma 2s \quad \sigma^* 2s \quad \sigma 2p \quad \pi_y 2p \quad \pi_y^* 2p \quad \sigma^* 2p$$
$$\pi_z 2p \quad \pi_z^* 2p$$

becomes

$$z\sigma \quad y\sigma \quad x\sigma \quad w\pi \quad v\pi \quad u\sigma.$$

$\sigma 1s$- and $\sigma^* 1s$-orbitals are called $(k)z\sigma$ and $(k)y\sigma$ whilst $\sigma 3s$- and $\sigma^* 3s$-orbitals are called $(m)z\sigma$ and $(m)y\sigma$. The (k) and the (m) denote the principal quantum number of the atomic orbitals normally concerned.

The disadvantages of the Mulliken notation are the possibilities of confusion between the x, y and z as used in the notation and as used in referring to Cartesian co-ordinates, and the lack of any direct notational link between a molecular orbital and the atomic orbitals from which it is generally formed and into which it generally reverts when the molecule is split up into atoms.

The advantage of the Mulliken system, apart from its brevity, occurs when the relation between molecular and atomic orbitals is not direct. A $z\sigma$-molecular orbital, for instance, need not originate from, or split into, $2s$- atomic orbitals. In hydrogen chloride, for example (page 99), a $z\sigma$-molecular orbital is derived from $1s$- and $2p_x$-atomic orbitals. A $z\sigma$-molecular orbital is simply the σ-molecular orbital of lowest energy.

9 Simple Examples of Covalent Bonding

In working out the arrangement of electrons in a simple molecule, the available electrons are assigned to the available molecular orbitals in energy order. The Pauli principle applies so that any molecular orbital can contain only two electrons, and these must have opposed spins. It is only when more electrons occupy bonding than antibonding orbitals that a chemical bond occurs.

The application of these ideas is easy enough if it is possible to work out the available molecular orbits in correct energy order. This is easier for homonuclear diatomic molecules than for heteronuclear ones, and it is still more difficult for polynuclear molecules.

HOMONUCLEAR DIATOMIC MOLECULES

1 Hydrogen molecule-ion, H_2^+ This is the simplest molecule that can exist, though it has no real chemical existence for it can only be detected spectroscopically when an electric discharge is passed through hydrogen gas under reduced pressure. In the molecule-ion, one electron holds two protons together. The single available electron occupies the $\sigma 1s$-bonding orbital and the structure of the molecule-ion is written as $H_2^+[(\sigma 1s)]$. The bond length is 0·106 nm.

2 Hydrogen, H_2 In the formation of the hydrogen molecule, H_2, the two $1s$-electrons, one from each of the atoms concerned, pass into a $\sigma 1s$-molecular orbital. This is a bonding orbital and constitutes the single covalent bond in the hydrogen molecule. Since the $\sigma 1s$-molecular orbital can only hold two electrons if they have different spins it is clear that a molecule will only be formed from two hydrogen atoms containing electrons with opposite spins. The σ^*1s-antibonding molecular orbital is unoccupied in the hydrogen molecule.

The structure of the hydrogen molecule is written as $H_2[(\sigma 1s)^2]$ or represented diagrammatically in what is known as a *molecular orbital diagram* or a *cell diagram* (Fig. 55). The bond length in the molecule is 0·074 nm.

3 Helium molecule-ion, He_2^+ This molecule-ion can be detected in a discharge tube. There are three available electrons. Two of them occupy a $\sigma 1s$-bonding orbital whilst the other is in the σ^*1s-anti-bonding orbital. The anti-bonding effect of one electron is greater than the bonding effect (page 88); the bond in the helium molecule-ion is, therefore, weaker than the single electron bond in the hydrogen molecule-ion. The structure is written as $He_2^+ [(\sigma 1s)^2(\sigma^* 1s)]$.

4 Helium The molecule, He_2, cannot exist under normal conditions, because two of the four available $1s$-electrons pass into a $\sigma 1s$-bonding orbital, whilst the other two pass into a $\sigma^* 1s$-anti-bonding orbital. The filled anti-bonding orbital more than counteracts the filled bonding orbital, so that no bond results; there is, in fact, some repulsion.

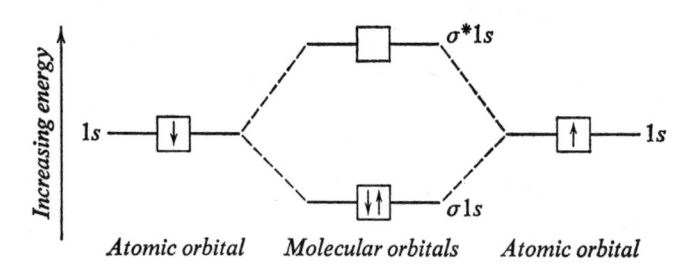

Fig. 55. The molecular orbital or cell diagram for the H_2 molecule

He_2 molecules can be detected in discharge tubes but they must be in an excited state with two or more electrons being promoted from the anti-bonding $\sigma^* 1s$-orbital to the bonding $\sigma 2s$-orbital.

5 Lithium, Li_2 Lithium vapour contains about 1 per cent of Li_2 molecules. The arrangement of electrons in an atom of lithium is $1s^2 2s^1$, and, as a first approximation, any bonding or non-bonding effect of electrons in inner shells is disregarded. These electrons in inner shells remain in their atomic orbitals and do not enter molecular orbitals.

Bonding in Li_2 molecules is caused, therefore, by the two available $2s$-electrons passing into a $\sigma 2s$-bonding molecular orbital. The resulting structure is written as $Li_2[KK(\sigma 2s)^2]$ the KK denoting the filled inner shells in the two lithium atoms.

Similarly the structure of Na_2 molecules is written as $Na_2[KKLL(\sigma 3s)^2]$.

6 Nitrogen, N_2 The way in which two N atoms combine to form a

N_2 molecule can be represented as in Fig. 56. The resulting structure for the N_2 molecule is then given by*

or
$$N_2[KK(\sigma 2s)^2(\sigma^* 2s)^2(\sigma 2p)^2(\pi_y 2p)^2(\pi_z 2p)^2]$$

$$N_2[KK(z\sigma)^2(y\sigma)^2(x\sigma)^2(w\pi)^4]$$

The bonding of electrons in the $\sigma 2s$-bonding orbital is neutralised by the anti-bonding of the electrons in the $\sigma^* 2s$-orbital, so that six

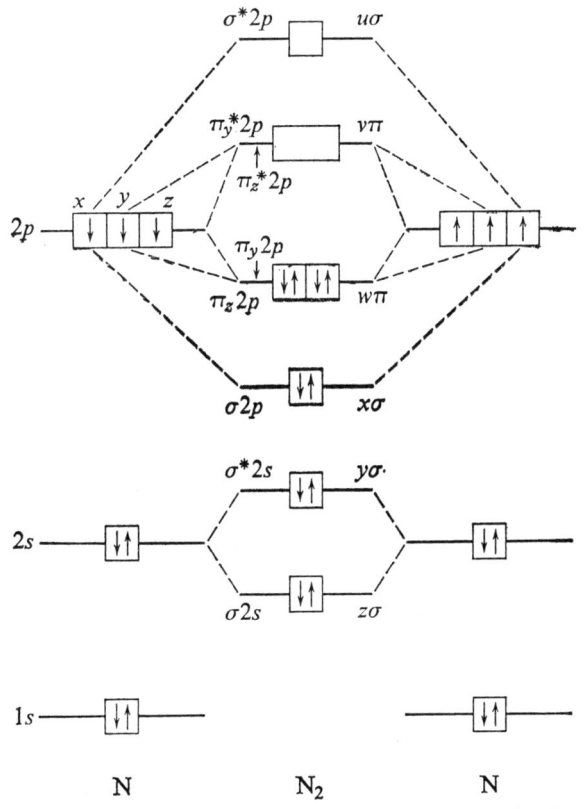

Fig. 56. Molecular orbital diagram for N_2

electrons remain in bonding orbitals. This represents a triple bond, written simply as N≡N, but one of the bonds is a σ-bond whilst two

* There is some evidence that the $x\sigma$ molecular orbital is of higher energy than the $w\pi$ in the nitrogen molecule. On this basis, the $(x\sigma)^2$ and $(w\pi)^4$, or the $(\sigma 2p)^2$ and $(\pi_y 2p)^2$ $(\pi_z 2p)^2$, terms in the structure as given must be interchanged.

are π-bonds. This is illustrated, diagrammatically, in Fig. 57. The bond is both very short (0·11 nm) and very strong (941 kJ mol^{-1}).

The similar structure of the P_2 molecule is written as

$$P_2[KKLL(\sigma 3s)^2(\sigma^* 3s)^2(\sigma 3p)^2(\pi_y 3p)^2(\pi_z 3p)^2]$$

7 Oxygen, O_2 The structure of the O_2 molecule is written as

$$O_2[KK(\sigma 2s)^2(\sigma^* 2s)^2(\sigma 2p)^2(\pi_y 2p)^2(\pi_z 2p)^2(\pi_y^* 2p)(\pi_z^* 2p)]$$

or

$$O_2[KK(z\sigma)^2(y\sigma)^2(x\sigma)^2(w\pi)^4(v\pi)^2]$$

and the way in which this is built up is shown in Fig. 58.

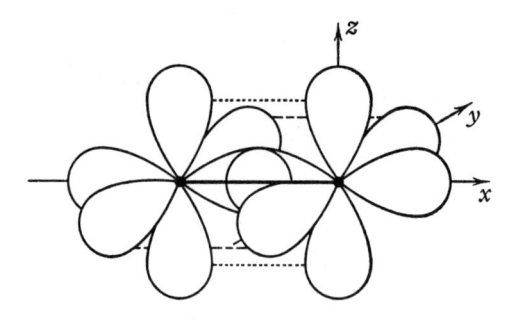

Fig. 57. Diagrammatic representation of the triple bonding in a N_2 molecule. The bold line shows the σ-bond formed by the two $2p_x$ orbitals. The dotted and dashed lines show the π-bonds formed by the $2p_z$ and $2p_y$ orbitals respectively

The lower molecular orbitals fill up quite normally, but there are only two electrons available for the $\pi_y^* 2p$- and $\pi_z^* 2p$-orbitals which, together, could hold four. If the two electrons were both allotted to either the $\pi_y^* 2p$- or the $\pi_z^* 2p$-orbital all electrons would be paired. The application of the rule of maximum multiplicity (page 27) shows, however, that one electron must be allotted to each orbital, and this means that the molecule contains two unpaired electrons.

These unpaired electrons account for the strong paramagnetism (page 64) of oxygen, and this explanation was one of the earliest successes of the application of molecular orbital theory.

There are four more electrons in bonding than in anti-bonding orbitals in the O_2 molecule. This represents a double bond, O=O, but one of the bonds is a σ-bond whilst the other is a π-bond. The bond length is 0·12 nm and the bond energy 498 kJ mol⁻¹.

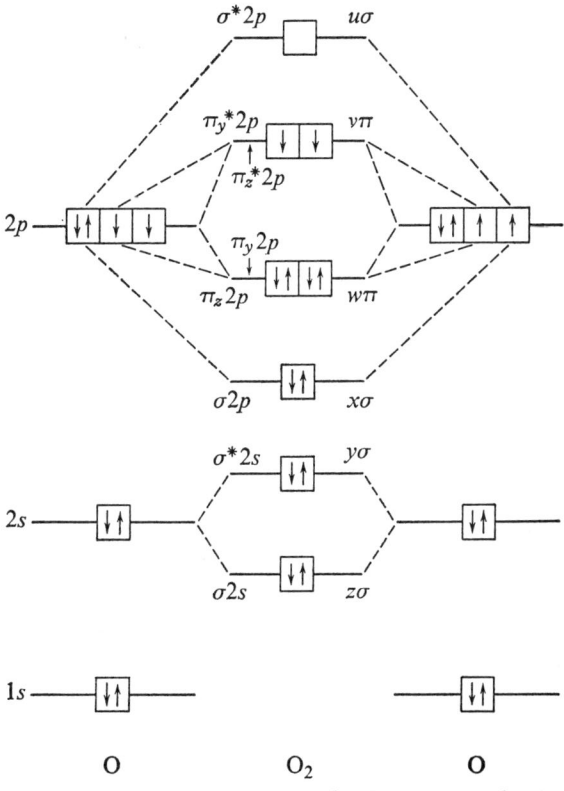

Fig. 58. Molecular orbital diagram for O_2

8 Fluorine, F₂ The F_2 molecule has two more electrons than the O_2 molecule. These occupy the π_y^*2p- and π_z^*2p-orbitals so that these two orbitals become fully occupied, giving a structure for fluorine of

$$F_2[KK(\sigma 2s)^2(\sigma^*2s)^2(\sigma 2p)^2(\pi_y 2p)^2(\pi_z 2p)^2(\pi_y^*2p)^2(\pi_z^*2p)^2]$$

or

$$F_2[KK(z\sigma)^2(y\sigma)^2(x\sigma)^2(w\pi)^4(v\pi)^4]$$

The only bonding orbital which is not neutralised by a corresponding anti-bonding orbital is the $\sigma 2p$-orbital. This constitutes the single σ-bond of the F_2 molecule. Chlorine, bromine and iodine molecules have precisely similar structures.

9 Summary Typical electronic structures for molecules of the light elements are summarised in Fig. 59, which does for simple molecules what Fig. 10 (page 27) does for simple atoms.

In each molecular species the excess of bonding over anti-bonding electrons, divided by two, gives the number of covalent bonds formed. Alternatively, this number is known as the bond order. It will be seen that the maximum number of covalent bonds that can be formed is three, for there are not enough bonding molecular orbitals for more bonds to be formed.

Figure 59 also shows that there is a rough proportionality, in some cases, between the number of covalent bonds formed and the dissociation energy of the molecule concerned. The dissociation energy is the energy necessary to split the molecule into two free atoms. This proportionality is to some extent accidental because the inter-nuclear distances in these molecular species are not the same. Moreover, the bonding is sometimes caused by 1s-electrons and sometimes by 2s- or 2p-electrons. Further details of bond energies are given on page 135.

HETERONUCLEAR DIATOMIC MOLECULES

The formation of molecular orbitals in homonuclear diatomic molecules is simplified by the fact that like atoms are being bonded. They must, necessarily, have atomic orbitals of equal energy and like symmetry, and, for the most part, any interaction other than that of ns orbital with ns, or np with np, can be neglected.

Such limiting factors do not, however, operate when two different atoms, A and B, combine to form a molecule, AB. The same general principles apply, but their application is less straightforward. In some cases, the iso-electronic principle is useful.

10 The iso-electronic principle Two molecular species with the same number of atoms and the same total number of valency electrons are said to be iso-electronic, and the iso-electronic principle states that such molecular species will have similar molecular orbitals and molecular structures. The molecular species involved may be molecules, anions or cations. The application of this principle enables some simple diatomic molecules to be likened to molecules of iso-electronic elements.

(a) *The nitrosyl cation, NO^+.* This ion is iso-electronic with nitrogen, N_2, both having two atoms and ten valency electrons. On this basis, the structure of NO^+ would be expected to be

$$NO^+[KK(\sigma 2s)^2(\sigma^* 2s)^2(\sigma 2p)^2(\pi_y 2p)^2(\pi_z 2p)^2]$$

	$\sigma 1s$	σ^*1s	$z\sigma$ $\sigma 2s$	$y\sigma$ σ^*2s	$x\sigma$ $\sigma 2p$	$w\pi$ $\pi_y 2p$ $\pi_x 2p$	$v\pi$ π_y^*2p π_x^*2p	$u\sigma$ σ^*2p	Bonding– anti-bonding electrons	Normal covalency	Bond order	Dissociation energy (kJ mol^{-1})
H_2^+	↑								1		½	256
H_2	↑↓								2	1	1	436
He_2^+	↑↓	↑							1		½	257
(He_2)	↑↓	↑↓							0	0	0	
Li_2	Not occupied (KK)		↑↓						2	1	1	110
N_2			↑↓	↑↓	↑↓	↑↓ ↑↓			6	3	3	941
O_2			↑↓	↑↓	↑↓	↑↓ ↑↓	↑ ↑		4	2	2	498
F_2			↑↓	↑↓	↑↓	↑↓ ↑↓	↑↓ ↑↓		2	1	1	158
$(Ne)_2$			↑↓	↑↓	↑↓	↑↓ ↑↓	↑↓ ↑↓	↑↓	0	0	0	

Fig. 59. The filling up of molecular orbitals in simple diatomic molecules

the bonding being a triple bond made up, as in nitrogen, of one σ-bond and two π-bonds. The molecular orbital diagram shown in Fig. 56, for nitrogen, would apply for NO$^+$, except that the energy of the atomic orbitals for the oxygen atom would be slightly lower than the corresponding atomic orbitals for nitrogen. This is because oxygen has a higher electronegativity (page 125) than nitrogen. The bond length in NO$^+$ is 0·106 nm, which compares with a value of 0·110 nm in nitrogen.

(b) *Nitrogen oxide, NO.* Nitrogen oxide has one more electron per molecule than NO$^+$ and its structure would be expected to be

$$NO[KK(\sigma 2s)^2(\sigma^* 2s)^2(\sigma 2p)^2(\pi_y 2p)^2(\pi_z 2p)^2(\pi_y^* 2p)]$$

The excess of bonding over anti-bonding electrons is 5, as compared with 6 in N$_2$ and NO$^+$. The bond-order of NO is, therefore, 2½, which represents a σ-bond and two π-bonds less the anti-bonding effect of an unpaired electron in the $\pi_y^* 2p$ molecular orbital. This single anti-bonding electron makes the bond energy in NO (678 kJ mol^{-1}) less than that of N$_2$ (941 kJ mol^{-1}). The bond length in NO is, correspondingly, longer; the value is 0·114 nm.

The unpaired electron in the nitrogen oxide molecule means that it is paramagnetic, like oxygen (page 94).

(c) *Carbon monoxide, CO.* Carbon monoxide is iso-electronic with nitrogen and, on this basis, it would be expected to have a structure written as

$$CO[KK(\sigma 2s)^2(\sigma^* 2s)^2(\sigma 2p)^2(\pi_y 2p)^2(\pi_z 2p)^2]$$

This represents triple bonding between the two atoms, made up of one σ-bond and two π-bonds.

Such a structure does not, however, account fully for all the facts. In the first place, carbon monoxide has a small dipole moment (page 126). Moreover, the bonding in CO is weaker than it is in CO$^+$. The loss of an electron in going from CO to CO$^+$ would be expected to occur from the $\pi_z 2p$ bonding orbital making the bond order in CO$^+$ 2½ as compared with that of 3 in carbon monoxide.

The matter cannot be resolved simply.

11 Hydrogen chloride In the formation of an HCl molecule it is only the 3p-electrons of the chlorine atom that can combine effectively with the hydrogen 1s-electrons, from the energy point of view. Combination of the hydrogen 1s-orbital with the $3p_y$- or $3p_z$-orbitals of the chlorine atoms is ruled out on symmetry grounds (page 86), so that only combination between H(1s) and Cl(3p$_x$) remains, to give a σ-bond between the atoms.

The molecular orbital formed by combination of these two atomic orbitals will not be symmetrical. The shape will be as shown in

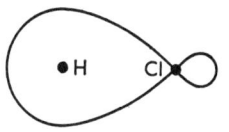

Fig. 60. Shape of molecular orbital formed by combination of H(*1s*) *and* Cl(*3p$_x$*) *atomic orbitals*

Fig. 60, and this indicates the polar nature of the bond, H$^{\delta+}$— Cl$^{\delta-}$. The lack of symmetry is caused by the fact that the Cl(3p$_x$) and the H(1s) atomic orbitals have not got the same energy. The molecular

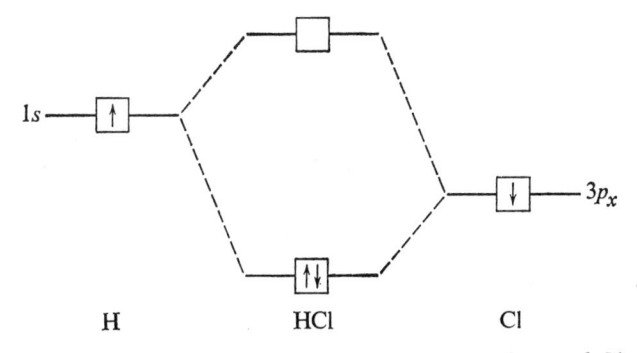

Fig. 61. The molecular orbital formed from H(*1s*) *and* Cl (*3p$_x$*) *atomic orbitals is closer in energy to the* Cl(*3p$_x$*) *than to the* H(*1s*).

orbital they form can be represented as in Fig. 61, and it will be seen that the molecular orbital has an energy closer to that of the Cl(3p$_z$) orbital than to that of the H(1s). This means that the molecular orbital resembles the atomic orbital of chlorine rather than that of hydrogen.

10 Directional Nature of Covalent Bonds

That covalent bonds are directed in space is shown both by the existence of stereoisomerism in covalent compounds (page 177) and by the fact that covalent molecules take up a wide variety of geometrical shapes. It is also possible to measure the actual bond angles between covalent bonds in molecules.

This directional nature of covalent bonding can be considered from a number of different points of view, which are outlined in this chapter.

DISTRIBUTION OF ELECTRON PAIRS AROUND A CENTRAL ATOM

1 Introduction A very useful correlation of the structures of many simple molecules and ions can be achieved by considering the way in which electron pairs might be expected to be distributed around a central non-transitional atom. The electron pairs concerned might be lone pairs or bonded pairs, i.e. pairs constituting a covalent bond. The electron pairs will repel each other so that they will take up positions as far apart from each other as possible. On the assumption that lone pairs cause a greater repulsion than bonded pairs, it is possible to account, qualitatively, for the bond angles found in a number of simple molecules and ions.

2 Distribution of two or three electron pairs An atom such as beryllium, with two valency electrons, will be surrounded by two bonded electron pairs in compounds such as beryllium chloride, $BeCl_2$. These electron pairs will repel each other as fully as possible so that the $BeCl_2$ molecule (page 108) will be linear.

Boron has three valency electrons and is, therefore, surrounded by three bonded electron pairs in all its BX_3 covalent compounds. As the pairs repel each other equally, the compounds, e.g. BCl_3 and $B(CH_3)_3$, have planar molecules with bond angles of 120°.

In tin(II) halides, e.g. $SnCl_2$, the central atom is surrounded by two bonded electron pairs and one lone pair. The bond angle is less than 120° because of the smaller repulsion between the two bonded pairs than between the bonded and the lone pairs.

3 Distribution of four electron pairs Four electron pairs would take up a tetrahedral arrangement around a central atom and this accounts for the tetrahedral arrangement to be found in methane, CH₄, and many other carbon compounds. Similar tetrahedral arrangements probably occur in covalent compounds of trivalent nitrogen, divalent oxygen and monovalent fluorine. The nitrogen atom in an ammonia, NH₃, molecule, for example, is surrounded by

Methane
Bond angle = 109° 28'
No lone pair

Ammonia
Bond angle = 107°
One lone pair

Water
Bond angles = 104° 31'
Two lone pairs

Hydrogen flouride
Three lone pairs

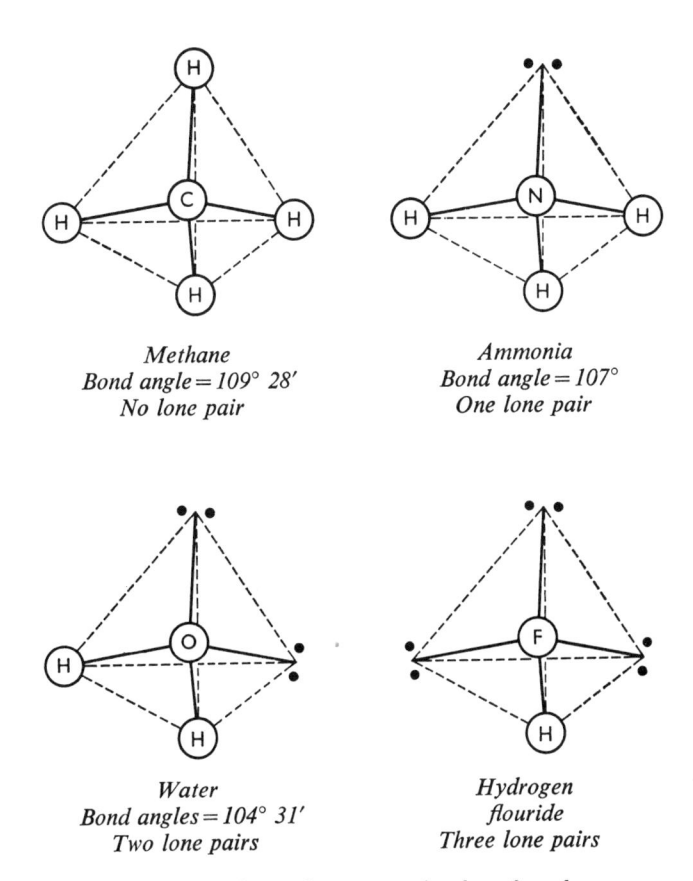

Fig. 62. The correlation between molecules of methane, ammonia, water and hydrogen fluoride. The tetrahedral distribution of four electron pairs around a central atom

three bonded electron pairs and one lone pair; oxygen in water is surrounded by two bonded pairs and two lone pairs; and fluorine in hydrogen fluoride is surrounded by one bonded pair and three lone pairs. The close relationship between these molecules and that of methane is shown in Fig. 62.

The bond angles in ammonia and water are less than the tetrahedral bond angle because of the greater repulsion effect of the lone pairs as compared with the bonding pairs.

The idea can also be extended to ions which are iso-electronic

PCl₅
No lone pairs

TeCl₄
One lone pair

ClF₃
Two lone pairs

ICl₂⁻
Three lone pairs

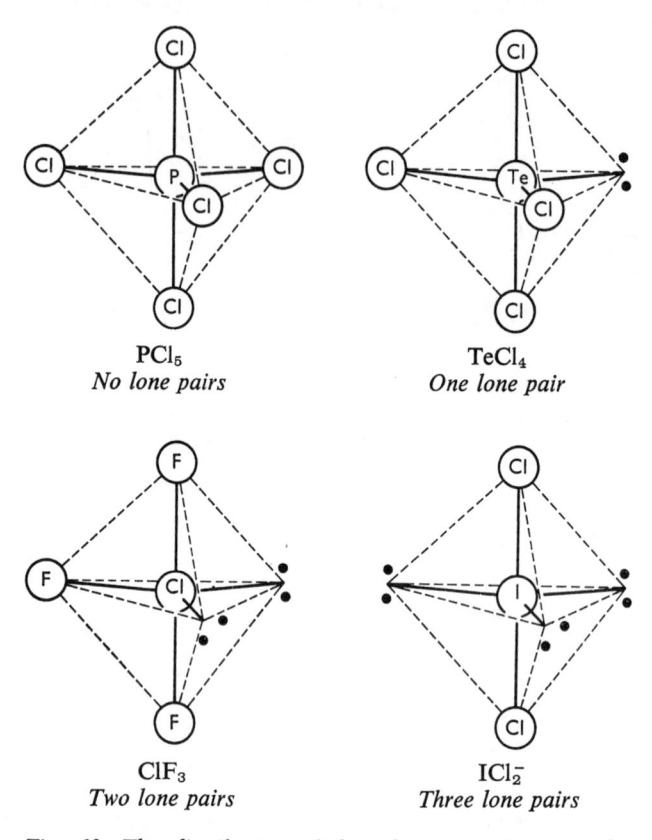

Fig. 63. The distribution of five electrons pairs around a central atom in a trigonal bi-pyramidal arrangement

(page 96) with carbon, nitrogen, oxygen or fluorine;

		$1s$	$2s$	$2p_x$	$2p_y$	$2p_z$	Typical ion
Like C	Be²⁻	↑↓	↓	↓	⋮	⋮	BeF₄²⁻
	B⁻	↑↓	↓	↓	↓	⋮	BF₄⁻
	N⁺	↑↓	↓	↓	↓	↓	NH₄⁺
Like N	O⁺	↑↓	↑↓	↓	↓	↓	H₃O⁺
Like O	N⁻	↑↓	↑↓	↑⋮	↓	↓	NH₂⁻
Like F	O⁻	↑↓	↑↓	↑↓	↑⋮	↓	OH⁻

The dotted arrows indicate electrons causing the ionic charge.

4 Distribution of five electron pairs Five electron pairs distribute themselves around a central atom in a trigonal bi-pyramidal arrangement. Such an arrangement occurs in phosphorus pentachloride, and other related structures involving lone pairs are shown in Fig. 63. The bond angles actually found in $TeCl_4$ and ClF_3 vary slightly from the bond angles shown in the regular geometrical shapes given in Fig. 63.

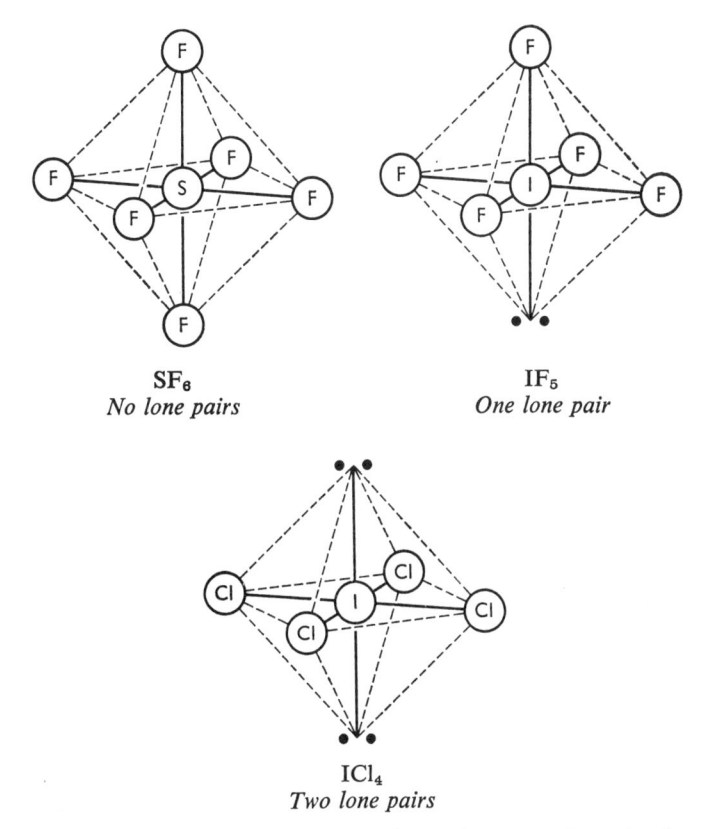

SF_6
No lone pairs

IF_5
One lone pair

ICl_4
Two lone pairs

Fig. 64. Octahedral arrangement of six electron pairs around a central atom

This can be attributed to the repulsion between lone pairs being greater than that between bonded pairs.

5 Distribution of six electron pairs Six electron pairs distribute themselves around a central atom octahedrally, as in sulphur hexafluoride. Other related structures involving lone pairs are shown in Fig. 64.

OVERLAP OF ATOMIC ORBITALS

6 s–p bonding One of the conditions under which atomic orbitals can be combined to form molecular orbitals is that the atomic orbitals must overlap (page 85). The overlapping atomic orbitals can only be combined if they also have similar energy and similar symmetry (page 85), but, taking these factors into account, the strength of a covalent bond is closely related to the extent of overlap of the atomic orbitals concerned. This is known as the *principle of maximum overlap* and it is extremely useful in a general consideration of covalent bonding in molecules.

Two s-orbitals cannot overlap very strongly because of the spherical distribution of charge. What might be called an s–s bond is, therefore, relatively weak.

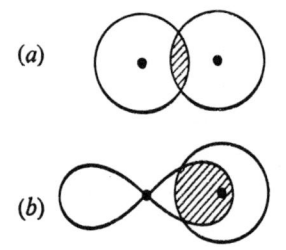

Fig. 65. The difference in overlap between (a) *two s orbitals and* (b) *one s and one p orbital*

Because p-orbitals are concentrated in a particular direction, and because their lobes are longer than the radius of the corresponding s-orbital, they can overlap with other s- or p-orbitals more effectively than two s-orbitals can overlap (Fig. 65). s–p σ-bonds are, therefore, stronger than s–s bonds, and p–p σ-bonds are stronger still. It can be shown that the relative bond strengths are of the order

s–s	s–p	p–p
1	1·732	3

Hydrogen halides (page 99) provide good examples of molecules in which s–p bonding plays an important part.

7 Directional nature of covalent bonds Because of the spherical symmetry of an s-orbital it is not concentrated in any particular direction. The three p-orbitals in a shell are, however, concentrated along the x-, y- and z-axes, mutually at right-angles to each other. d- and f-orbitals also have a directional nature.

It is the directional nature of p-, d- and f-orbitals, and of hybrid orbitals (page 107), which accounts for the directional nature of the covalent bond.

(a) *Water, H_2O.* The two O—H bonds in a water molecule are, essentially, s–p bonds like the bond in the hydrogen halides (page 99) and the formation of the molecule from one oxygen and two hydrogen atoms can be represented as in Fig. 66.

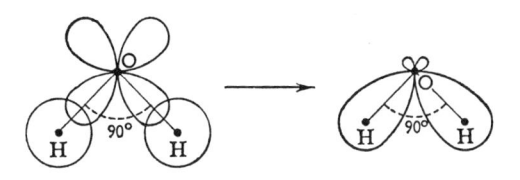

Fig. 66. The way in which two s orbitals from hydrogen atoms might be expected to interact with p orbitals from an oxygen atom to form a water molecule with a bond angle of 90°

If one of the O—H bonds is along the x-axis and the other along the y-axis, the $2p_x$-atomic orbital of the oxygen atom will be able to combine with a $1s$-atomic orbital of one hydrogen atom to form a σ-orbital, whilst the $2p_y$-orbital of the oxygen atom will combine, similarly, with a $1s$-orbital of the other hydrogen atom.

The bond angle in such a molecule would be expected to be 90° but the actual bond angles found in water and other similar molecules are:

H_2O	H_2S	H_2Se	H_2Te
104°31′	93·3°	91°	90°

The fact that the actual bond angles are greater than 90° is explainable in terms of some sp-hybridisation (page 110) and the ionic character (page 122) of the O—H bond which causes a repulsion between the two positively charged hydrogen atoms (Fig. 67a).

(b) *Ammonia, NH_3.* A nitrogen atom, with a structure of $1s^2 2s^2 2p^3$ has three p-electrons available for bond formation. As the three p-orbitals are mutually at right-angles they should form s–p bonds by overlap with $1s$-atomic orbitals of three hydrogen atoms so that the resulting molecule of ammonia, NH_3, will be pyramidal with the N—H bonds at 90° to each other (Fig. 67b).

The actual bond angles in ammonia and similar molecules are as follows:

| NH$_3$ | PH$_3$ | AsH$_3$ | SbH$_3$ |
| 107·3° | 93·3° | 91·8° | 91·3° |

and the discrepancy between these values and 90° can be accounted for in terms of ionic character causing repulsion between the hydrogen atoms, and by hybridisation, as for water.

The non-existence of optical isomers of substituted ammonias of the type Nabc is accounted for by the fact that the molecule easily turns itself inside-out, umbrella-like.

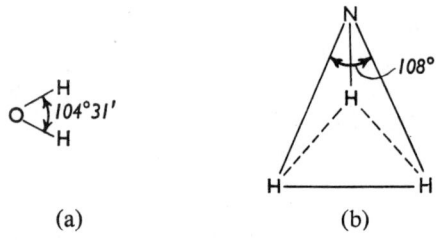

(a)   (b)

Fig. 67. Bond angles in (a) water, and (b) ammonia molecules

HYBRIDISATION

8 Promotion of electrons The formation of bonds by the overlap of two atomic orbitals (page 104), requires that each of the two orbitals must contain one unpaired electron and that the two electrons must have different spins. This is because the molecular orbital formed by the overlap can only hold two electrons and they must have opposed spins by the Pauli principle (page 16).

In this way, then, the number of bonds which an atom might be able to form depends, to some extent, on the number of unpaired electrons in the atom. A study of the electronic arrangements in the atoms of atomic number 1 to 10, given on page 27, shows that the number of unpaired electrons is, in fact, equal to the commonest numerical covalency of the element concerned in all cases except beryllium, boron and carbon.

The significant data are repeated below:

Element	H	He	Li	Be	B	C	N	O	F	Ne
Number of unpaired electrons	1	0	1	0	1	2	3	2	1	0
Numerical valency	1	0	1	2	3	4	3	2	1	0

In terms of unpaired electrons, beryllium would be expected to behave as an inert gas, boron might be expected to be monovalent like fluorine, and carbon would be divalent like oxygen.

	$1s$	$2s$	$2p_x$	$2p_y$	$2p_z$
Be	↑↓	↑↓			
B	↑↓	↑↓	↓		
C	↑↓	↑↓	↓	↓	

To meet the known chemical facts it must be assumed that some of the paired electrons are uncoupled or unpaired before the atom participates in chemical bonding. This will require an input of energy, but such energy will be available from the heat of reaction when covalent bonds are formed.

The unpairing of electrons requires the removal of a $2s$-electron into a $2p$-orbital of higher energy level. This process is referred to as *promotion*, and the arrangement of electrons after promotion is sometimes referred to as an *excited valency state*.

The simplest excited valency states for beryllium, boron and carbon are as follows:

	$1s$	$2s$	$2p_x$	$2p_y$	$2p_z$
Be*	↓↑	↓	↓		
B*	↓↑	↓	↓	↓	
C*	↓↑	↓	↓	↓	↓

Such excited valency states, generally denoted by an asterisk, provide enough unpaired electrons for beryllium to form two, boron three, and carbon four, bonds, but they are not the bonds actually found in practice.

The excited structure given for carbon would give three bonds involving p-orbitals which would be at right-angles to each other together with a single bond involving an s-orbital. Methane, CH_4, for example, would be expected, by comparison with water and ammonia, to contain three s–p bonds mutually at right-angles, and one, weaker, s–s bond. It is well established, however, that the four bonds in methane are all alike and are arranged tetrahedrally.

Similarly, the bonds in boron trichloride, BCl_3, are all alike, and the molecule is planar with bond angles of 120° which would not be expected if the boron atom made use of two p- and one s-orbital separately.

9 Hybrid orbitals The answer to the problems posed at the end of the preceding section lies in the concept of hybrid orbitals. The distinction between s- and p- and d-orbitals must be abandoned, so long as they have comparable energies, and it must be assumed that these orbitals can be combined in various ways within an atom to form different, but equivalent, hybrid orbitals. As will be seen, such hybrid orbitals are more effective in forming strong bonds than 'pure' s-, p- and d-orbitals (page 110).

Hybridisation, or mixing, is simply a matter of taking s- and/or p- and/or d-atomic orbitals and combining their wave functions either by adding or subtracting, to give new wave functions representing new hybrid orbitals.

In the following sections, hybridisation of s- and p-orbitals will be considered first. Hybridisation involving d-orbitals is discussed on pages 112–13, and the important effect of hybridisation on complex ions is dealt with on pages 165–9.

10 sp (Linear or digonal) hybridisation The combination of an s- and a p-orbital leads to two hybrid orbitals known as sp-orbitals. One is

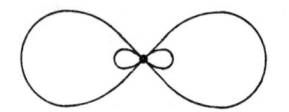

Fig. 68. Two co-linear sp orbitals

obtained by adding the two wave functions, and the other by subtracting them. The two sp-orbitals are co-linear, when superimposed, as in Fig. 68.

The large lobes of the two sp-orbitals protrude along the axis further than the corresponding s- or p-orbitals. A hybrid sp-orbital is, therefore, able to form a stronger bond, by overlap with another orbital of another atom, than either s- or p-orbitals alone (see page 104). It is only, then, by using these hybrid orbitals that the strongest bonds and the maximum stability can be attained.

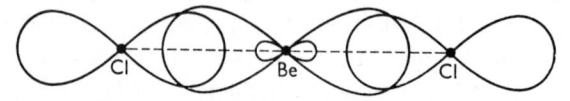

Fig. 69. The formation of two σ-bonds in $BeCl_2$ by overlap of $3p_x$ orbitals from two chlorine atoms with two hybrid sp orbitals from a beryllium atom

(a) *Beryllium chloride, $BeCl_2$.* Solid beryllium chloride is polymerised and exists as $(BeCl_2)_2$ but it is likely that individual co-linear $BeCl_2$ molecules exist in solution and in the vapour state.

The two sp-orbitals of beryllium overlap two $3p_x$-orbitals from two chlorine atoms to form two σ-bonds as illustrated in Fig. 69.

(b) *Ethyne*, C_2H_2. In the ethyne molecule, hybridisation of one $2s$- and one $2p$-carbon orbitals leads to each of the two carbon atoms having co-linear *sp*-hybrid orbitals. These overlap to form a σ-bond between the two carbon atoms and each of the two *sp*-orbitals also overlaps with a $1s$-atomic orbital of a hydrogen atom to form σ-bonds between the carbon and hydrogen atoms. It is because *sp*-hybrid orbitals are concerned that the ethyne molecule is co-linear (Fig. 70).

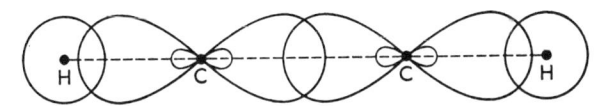

Fig. 70. σ-bonds in ethyne formed by overlap of two sp hybrid orbitals (between the two carbon atoms) and by overlap of an sp hybrid and an s orbital (between the carbon and hydrogen atoms)

Each of the carbon atoms has two remaining $2p$-atomic orbitals which interact to form two π-bonds between the carbon atoms, these two bonds being in planes at right-angles to each other. The C≡C bond consists, therefore, of a σ-bond and two π-bonds as in Fig. 71.

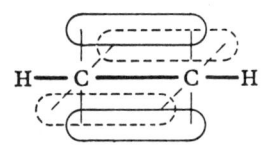

Fig. 71. The bonding in an ethyne molecule. Two σ-bonds between carbon and hydrogen atoms. One σ-bond and two π-bonds (in planes at right angles to each other) between carbon atoms

(c) *sp Hybridisation in hydrogen halides*. Because *sp*-hybrid orbitals can form stronger bonds than either *s*- or *p*-orbitals alone it is likely that some measure of *sp*-hybridisation occurs in many molecules often regarded as forming bonds using *s*- and *p*-orbitals alone.

The main bonding in hydrogen fluoride and other hydrogen halides (page 99) is, for example, generally regarded as formed by the overlap of a $1s$-atomic orbital of hydrogen and a $2p_x$-atomic orbital of fluorine.

A better structure involves a $1s$-hydrogen orbital with a fluorine orbital which is mainly $2p_x$ but partially $2s$. Such a hybrid fluorine orbital would give a stronger H—F bond and also accounts more effectively for the measured dipole moment of hydrogen fluoride. It does this because the distribution of the lone pairs of the fluorine atom are influenced by the *sp*-hybridisation as is illustrated in Fig. 72. The effect of this is to place one lone pair in an orbital mainly beyond the F atom on the H—F axis away from the hydrogen atom. Such an arrangement contributes substantially to a dipole moment in the hydrogen fluoride molecule.

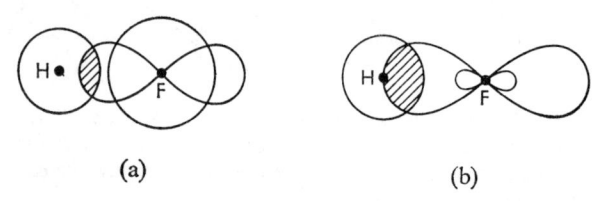

Fig. 72. Bonding in HF. (a) *Overlap of hydrogen 1s and fluorine* $2p_x$ *orbitals.* (b) *Overlap of hydrogen 1s and fluorine sp hybrid orbital. The* $2p_y$ *and* $2p_z$ *orbitals are not shown*

(*d*) *sp Hybridisation in water and ammonia.* The simple picture of the water molecule as involving *s–p* bonding between oxygen and hydrogen atoms ought to give a bond angle of 90°, whereas the actual bond angle is more than this (page 106).

If the two *s–p* bonds are, in fact, made from *s*-orbitals of hydrogen and oxygen hybrid orbitals consisting mainly of 2*p*- but partially of 2*s*-orbitals a bond angle greater than 90° would be expected. In the extreme, if oxygen could form fully hybridised *sp*-orbitals the H$_2$O molecule would be linear like beryllium chloride (page 108).

Some measure of *sp*-hybridisation can, therefore, account for the bond angle in water, and similarly in ammonia, but an alternative viewpoint is possible as explained on page 101.

11 sp^2 **(Trigonal) hybridisation** Combination of one *s*- and two *p*-orbitals gives rise to three sp^2-hybrid orbital which are co-planar and directed at angles of 120° to each other (Fig. 73). As with *sp*-hybrid orbitals, the sp^2 hybrids are able to form stronger bonds than *s*- or *p*-orbitals alone because, on account of their size, they can overlap other orbitals more effectively.

(*a*) *Boron trichloride,* BCl_3. Boron trichloride is a very volatile liquid. Its molecule is planar with bond angles of 120°. Its formation is due to the overlap of three sp^2-orbitals of boron with three $3p$-orbitals from each of three chlorine atoms.

Fig. 73. Three sp^2 orbitals in the same plane but at angles of 120° to each other

The other halides of boron have similar structures as have many alkyls, e.g. $B(CH_3)_3$ and $B(C_6H_5)_3$, and the borate ion, BO_3^{3-}.

(b) *Ethene, C_2H_4*. In ethene, each of the two carbon atoms form bonds through sp^2-orbitals. Two of the orbitals of each atom form

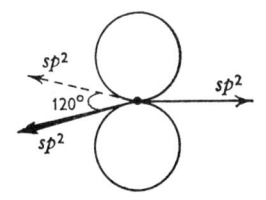

Fig. 74. The 2p orbital remaining on a carbon atom after the formation of three coplanar sp^2 orbitals

σ-bonds with the $1s$-orbitals of hydrogen atoms. The remaining sp^2-orbital of each carbon atom forms a σ-bond between the carbon atoms. The two carbon and four hydrogen atoms are all in the same plane and the bond angles are 120° (Fig. 74).

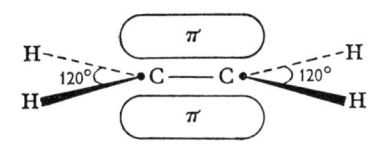

Fig. 75. The bonding in an ethene molecule. Four σ-bonds between carbon and hydrogen atoms. One σ-bond and one π-bond between carbon atoms

At right-angles to this plane there remains the unchanged $2p$-orbital of each carbon atom (Fig. 75), and these two $2p$-orbitals interact to form a π-bond between the two carbon atoms. The double bond between the carbon atoms consists, therefore, of a σ-bond and a π-bond.

12 sp^3 (Tetrahedral) hybridisation The orbitals formed by hybridisation of one s- and three p-orbitals are known as sp^3-orbitals; they are directed towards the corners of a tetrahedron. The bond angles in the resulting tetrahedral molecules are always close to the expected theoretical angle of 109° 28'.

In methane, CH_4, for example, $1s$-orbitals of each hydrogen atom overlap with an sp^3-hybrid orbital of the carbon atom; the bond angle in methane is 109° 28'. Other similar, symmetrical molecules involving sp^3-hybrid orbitals include the tetrahalides of carbon and silicon, SiH_4, $SnCl_4$ and $SnBr_4$, and $Pb(C_2H_5)_4$. Ions with tetrahedral arrangements include SO_4^{2-}, ClO_4^-, NH_4^+, BH_4^- and BF_4^- (see, also, page 102).

If some of the hydrogen atoms in a methane molecule are replaced by other atoms or groups the resulting molecule will be asymmetrical. Some deviation from the perfect tetrahedral arrangement might not be unexpected because of steric effects and because some of the bonds may have considerable ionic character (page 122). The deviations actually found are not, however, large. In dichloromethane, for example, the ClCCl angle is about 111°, and, in trichloromethane, it is about 112°.

13 d^2sp^3 or sp^3d^2 (Octahedral) hybridisation A combination of one s-, three p- and two d-atomic orbitals leads to six hybrid orbitals which are directed octahedrally, i.e. away from the origin in the $\pm x$, $\pm y$, and $\pm z$ directions (Fig. 105).

Sulphur hexafluoride provides an example. It is a colourless, odourless, inert gas made by burning sulphur in fluorine.

The ground-state arrangement of electrons in the sulphur atom is shown at (a), below

	$1s$	$2s$	$2p_x$	$2p_y$	$2p_z$	$3s$	$3p_x$	$3p_y$	$3p_z$	$3d$
(a) S	↑↓	↑↓	↑↓	↑↓	↑↓	↑↓	↑↓	↓	↓	
(b) S**	↑↓	↑↓	↑↓	↑↓	↑↓	↓	↓	↓	↓	↓ ↓

Before sulphur can form octahedral bonds by sp^3d^2 hybridisation a double promotion is necessary to give the excited state shown at (b). sp^3d^2-hybrid orbitals can then be formed and will be directed octahedrally. By overlap with $2p$-orbitals of fluorine atoms the octahedral SF_6 molecule is formed.†

† In more detailed discussion of this molecule the possible ionic nature of the S—F bonds must be taken into account.

Oxygen hexafluoride does not exist because there are no 2d-atomic orbitals which could make sp^3d^2 hybridisation possible. Similarly, SiF_6^{2-} and PF_6^- can be formed but not CF_6^{2-} or NF_6^-. This sort of fact used to be explained in terms of a covalency maximum which related the maximum number of covalent bonds an element could form to its position in the periodic table, as shown in the following table:

Atomic number	1	3–9	11–35	37–92
Element	H	Li–F	Na–Br	Rb–U
Maximum covalency	1	4	6	8

The d-orbitals involved in sp^3d^2 hybridisation, as in sulphur hexafluoride, originate from the same shell as the s- and p-orbitals. Alternatively, the d-orbitals involved may originate from a lower shell than the s- and p-orbitals; the hybrids are then called d^2sp^3-orbitals (see page 166).

14 sp^3d (Trigonal bi-pyramidal) hybridisation sp^3d-hybrid orbitals are directed towards the corners of a trigonal bi-pyramid, as shown in Fig. 102. They form three co-planar bonds with bond angles of 120° and two bonds at right-angles to this plane, one above and one below.

Phosphorus pentachloride, in the vapour state in which it exists as PCl_5 molecules, provides a common example of a compound involving sp^3d hybridisation. Phosphorus has the structure shown at (a)

		1s	2s	2p	3s	3p	3d
(a)	P	2	2	222	2	111	
(b)	P*	2	2	222	1	111	1
						sp^3d hybrid	

and this, by the promotion of an electron, becomes the excited valency state given at (b). This structure provides the opportunity for sp^3d hybridisation, the hybrid orbitals overlapping with the 3p-orbitals of five chlorine atoms.

NCl_5 could not exist, as predicted by the rules of maximum covalency (above), because 2d-orbitals do not exist so that sp^3d hybridisation is impossible for a nitrogen atom.

15 Summary of hybrid orbitals The main types of hybridisation which occur, together with some typical examples, are summarised in the table below:

Hybrid	No. of orbital bonds	Geometrical arrangement		Typical examples		
sp	2	Linear or digonal		$BeCl_2$	C_2H_2	
sp^2	3	Trigonal		BCl_3	C_2H_4	BO_3^{3-}
sp^3	4	Tetrahedral		CH_4 SO_4^{2-}	CCl_4 NH_4^+	$SnCl_4$ $Zn(NH_3)_4^{2+}$
dsp^2	4	Square		$Ni(CN)_4^{2-}$		$Cu(NH_3)_4^{2+}$
sp^3d	5	Trigonal bi-pyramidal		PCl_5		
d^2sp^3	6	Octahedral		$Fe(CN)_6^{3-}$ $Cr(NH_3)_6^{3+}$		$Fe(CN)_6^{4-}$
sp^3d^2	6	Octahedral		SF_6 FeF_6^{3-}	UF_6 $Cu(NH_3)_6^+$	SeF_6

DELOCALISED ORBITALS

16 Introduction In making use of molecular orbitals it is often necessary to limit the number of orbitals considered so as to maintain the conception of the chemical bond. In any polyatomic molecule, the molecular orbitals really belong to the molecule as a whole and not just to a particular bond within the molecule. The idea of specific bonds between pairs of atoms in a molecule can be maintained, however, by considering only the important molecular orbits which can be built up from pairs of atomic orbitals, or hybrid orbitals, associated with adjacent atoms. Such molecular orbitals, which can be regarded as bonding two particular atoms in a molecule, are known as *localised orbitals*.

In some compounds, however, it is not possible to treat the matter entirely in terms of localised orbitals. Some orbitals which must be considered must be regarded as belonging to the molecule as a whole and not just to a pair of atoms. Such orbitals are known as delocalised or non-localised orbitals. Their nature will be clarified by a consideration of the structures of buta-1, 3-diene and benzene.

17 Buta-1,3-diene Buta-1,3-diene is a gas which can be obtained by the catalytic dehydrogenation of butane by passing over a hot catalyst such as chromium(III) oxide. Its simple structural formula is shown, but such a structure is not satisfactory for it cannot account for the fact that the addition products from the reaction between buta-1,3-diene and bromine are 3.4 dibromobut-1-ene *and* 1.4 dibromobut-1-ene.

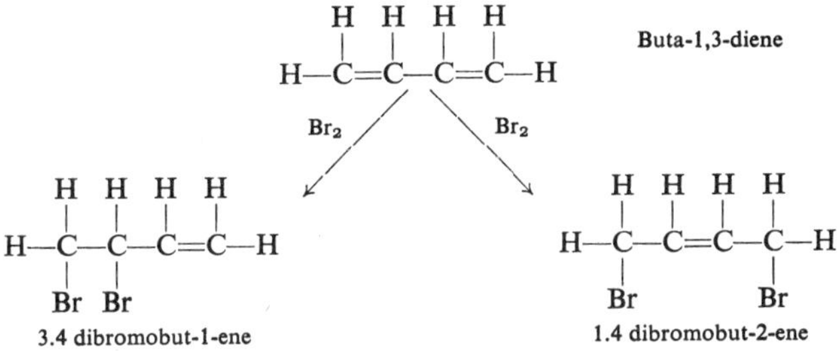

```
            H   H   H   H
            |   |   |   |          Buta-1,3-diene
        H—C=C—C=C—H
           Br₂ /    \ Br₂
```

```
    H  H  H  H                      H  H  H  H
    |  |  |  |                      |  |  |  |
  H—C—C—C=C—H                   H—C—C=C—C—H
    |  |                                |     |
    Br Br                               Br    Br
  3.4 dibromobut-1-ene            1.4 dibromobut-2-ene
```

A more satisfactory representation of the structure of buta-1,3-diene involves the introduction of de-localised orbitals.

The molecule is planar with bond angles of 120°. This is because the carbon atoms form σ-bonds with hydrogen, or other carbon, atoms by sp^2 hybridisation, just as the carbon atoms in an ethene molecule (page 111) do. The σ-bonds in the molecule are shown in Fig. 76.

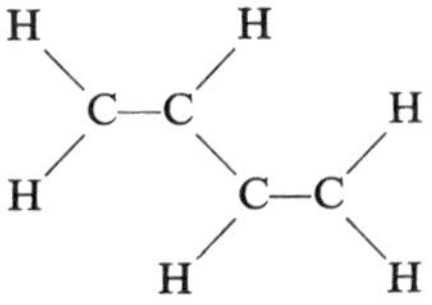

```
       H         H
        \       /
         C—C        H
        /    \     /
       H      C—C
             /    \
            H      H
```

Fig. 76. σ-bonds in buta-1,3-diene

Each carbon atom still retains a *p*-orbital and these four *p*-orbitals can overlap, as in ethene, to form π-bonds. The overlap of the *p*-orbitals cannot, however, be limited to pairs of adjacent atoms. The *p*-orbital of C_3 can interact with that of C_2 as readily as with that of C_4. The π-orbitals formed are not, therefore, associated with any particular bond in the molecule. They are, instead, spread out over the whole length of the molecule, both above and below it (Fig. 77); they are de-localised orbitals.

Each bond between carbon atoms is, in fact, a σ-bond with a certain amount of π-bonding.

18 Benzene The molecular orbital treatment of the benzene molecule envisages the electrons of each carbon atom in the molecule as existing in a state of sp^2-hybrid orbitals. Each carbon atom will, therefore, form three co-planar sp^2 σ-bonds directed at angles of 120° to each other (Fig. 78). Each carbon atom still holds a *p*-orbital at

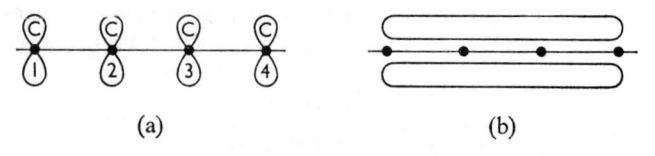

Fig. 77. (a) The four p-orbitals on the four carbon atoms in a buta-1,3-diene molecule. (b) The de-localised π-orbital formed from the four p-orbitals

right-angles to the plane of the carbon atoms (Fig. 79), and these six *p*-orbitals form de-localised π-orbitals pictured as charge-clouds all around the ring, both above and below it (Fig. 80).

Any system containing a —C=C—C=C— arrangement of bonds is known as a *conjugated* system. In all such systems, whether they be

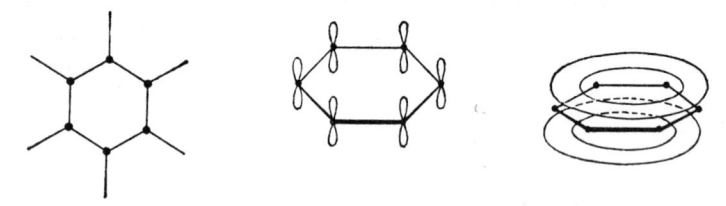

Fig. 78 (left). The σ-bonds in a benzene molecule. Fig. 79 (centre). The six p-orbitals at right angles to the plane of the carbon atoms in a benzene molecule. Fig. 80 (right). Formation of non-localised π-orbitals all round the benzene ring

ring systems, as in benzene, or straight-chain systems, as in buta-1,3-diene, the formation of de-localised orbitals greatly affects the bond lengths, the bond energies and the chemical properties of the molecule concerned.

11 Resonance and Electronegativity

1 Resonance hybrids The actual arrangement of electrons in a particular molecule cannot always be satisfactorily represented, using accepted symbols, solely in terms of simple ionic or covalent bonds. No single, simple structural formula which can be written for benzene (page 116), for example, accounts for all the known properties of benzene, and this is so for many compounds.

There is, in fact, plenty of evidence that the actual chemical bonds occurring in chemical compounds are of a type intermediate between pure ionic and pure covalent bonds. A covalent bond may have some ionic character, or a bond may be something in between a single and a double covalent bond.

Such ideas can be expressed in terms of molecular orbitals (page 99), but before the development of molecular orbital theory recourse had to be made to the conception of resonance, introduced and developed mainly by Pauling.

The actual electronic arrangement in a compound, which cannot be satisfactorily represented in one single structural formula using accepted symbols, had to be represented in terms of other possible, though non-existent, structures which could be formulated using accepted symbols. The actual structure which must be visualised does not consist of a mixture of the various possible structures. It is a single structure of its own, but as it cannot be written down simply on paper it is convenient to think of it in terms of structures which can.

A carbon dioxide molecule, for instance, can be represented by the three possible structures shown below:

I. $O=C=O$ or (dot structure)

II. $O \leftarrow C \ O$ or (dot structure)

III. $O \rightleftharpoons C \rightarrow O$ or (dot structure)

The actual structure of carbon dioxide which best accounts for all its properties, particularly for the measured bond lengths and heat of formation (page 119), must be considered as something closely related to all the three possible structures, but something which is different from all of them.

The actual structure is said to resonate between the structures I, II and III, or to be a *resonance hybrid* of the three structures. The individual structures, I, II and III, are known as *canonical forms*. Alternatively, using a terminology introduced by Ingold, the term mesomeric forms or mesomeric structures are used, and the conception is known as *mesomerism* ('between the parts').

Various analogies have been suggested to facilitate the building up of a visual picture of what resonance means. Perhaps the best is the idea of describing a rhinoceros in terms of a unicorn and a dragon. The rhinoceros, which has an actual existence, is thought of, as a sort of resonance hybrid, in terms of unicorns and dragons, which do not exist.

It must be emphasised that all the molecules in a resonance hybrid are alike, and that it is not just a mixture of different molecules. It is particularly important to distinguish between resonance and tautomerism, for they are easily confused. The latter may be regarded as the existence of two or more forms of a substance having different arrangements of *atoms*; the forms can sometimes be isolated. The possible structures contributing to a resonance hybrid have the same arrangement of atoms but different arrangements of electrons; they can never be separated for they have no real existence.

2 Resonance energy One of the most important points connected with resonance is that the resonance hybrid is a more stable structure than any of the structures contributing to it.

The increased stability is accounted for in the following way. If it is supposed that a resonance hybrid has two resonating structures, I and II, they can each be represented by wave functions, in simple cases, which give the energies of the structures they represent. By combining the two wave functions, it is found that a third wave function will also represent the system and that this function corresponds to a lower energy value, i.e. a higher stability.

The resonance energy of a substance is the extra stability of the resonance hybrid as compared with the most stable of the resonating structures (Fig. 81). It will be seen that the resonance hybrid is more stable than any of the resonating structures. It is, therefore, wrong to speak of the resonance hybrid being intermediate between the resonating structures so far as energy is concerned.

For carbon dioxide, the resonance energy is 154 kJ mol^{-1}. This value is obtained by subtracting the calculated heat of formation of

```
Increase in Stability →
Increase in Energy →
```
─────────── Energy of structure I

─────────── Energy of structure II
            ↑
     Resonance Energy
            ↓
─────────── Energy of actual structure

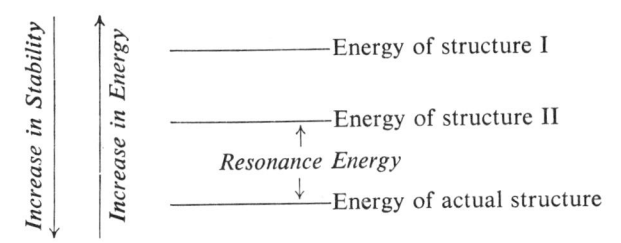

Fig. 81. Illustration of the increased stability of the actual structure caused by resonance between two possible structures, I and II

the O=C=O structure (1448 kJ, page 137) from the observed value of the heat of formation (1602 kJ). In general

$$\text{Resonance energy} = \text{Observed heat of formation} - \text{Calculated heat of formation}$$

3 Conditions necessary for resonance The several canonical forms which may contribute to a resonance hybrid cannot be chosen at random. They must all have the various atoms in the same relative positions; change in the position of the atoms leads to tautomerism. The several structures must also be of comparable energies; if very unstable, i.e. if of relatively high energy, the contribution they would make to the resonance hybrid would be negligible. Finally, the number of unpaired electrons must be the same in all the several structures to allow a continuous change from one bond type to another.

4 Examples of resonance Some typical examples of substances which can be represented as resonance hybrids are given below:

(a) *Carbon dioxide*. The formula of carbon dioxide was, for a long time, thought to be O=C=O. If this was the correct formula the bond distances in carbon dioxide ought to be equal to the sum of the double-bond covalent radii of carbon (0·067 nm) and oxygen (0·055 nm). This calculated bond length of 0·122 nm is not in agreement with the measured value of 0·115 nm.

Moreover, the bond energy of the C=O bond is 724 kJ mol^{-1} so that the calculated heat of formation of carbon dioxide, if its structure is O=C=O, would be 1448 kJ mol^{-1}. The actual measured value is 1602 kJ mol^{-1}.

It is, therefore, much better to represent carbon dioxide as a resonance hybrid between the possible structures given on page 117.

The structures I and II must contribute equally for they would, individually, have dipole moments whereas carbon dioxide has no dipole moment. The resonance energy (page 118) of carbon dioxide is

154 kJ mol^{-1}, i.e. the difference between the measured and calculated heats of formation.

(b) *Carbon monoxide.* The electronic structure to be allotted to carbon monoxide, with carbon normally four-valent and oxygen two-valent has always provided something of a problem. It is best represented as

$$\underset{\text{I}}{C=O} \longleftrightarrow \underset{\text{II}}{C \leftarrow O}$$

The calculated bond distances for structures I and II are 0·122 nm and 0·110 nm respectively, whereas the measured bond length is 0·113 nm.

Carbon monoxide has a dipole moment which is very nearly zero whereas both structure I and structure II would have large dipole moments if they existed individually.

The heat of formation of the C=O bond is 724 kJ mol^{-1}, whereas the observed heat of formation for carbon monoxide is 1071 kJ mol$^-$, the resonance energy being 347 kJ mol^{-1}.

(c) *Dinitrogen oxide.* A resonance hybrid between

$$\underset{\text{I}}{N=N=O} \longleftrightarrow \underset{\text{II}}{N\equiv N \rightarrow O}$$

The calculated bond lengths for the various bonds involved in these structures are N=N, 0·12 nm; N≡N, 0·11 nm, N=O, 0·115 nm; and N—O, 0·136 nm. The length of the molecule, which is linear, is 0·231 nm, the probable bond lengths being N—N, 0·112 nm and N—O, 0·119 nm.

The dipole moment of nitrous oxide is very small so that the two structures must contribute almost equally.

(d) *Nitrogen oxide.* A resonance hybrid between

$$\overset{x}{\underset{x}{N}}\overset{\bullet\bullet}{\underset{\bullet\bullet}{\vdots}} \overset{\bullet\bullet}{\underset{\bullet\bullet}{O}}\overset{\bullet}{\underset{\bullet}{\vdots}} \longleftrightarrow \overset{x\bullet}{\underset{x}{N}}\overset{\bullet}{\underset{\bullet\bullet}{\vdots}} \overset{\bullet\bullet}{\underset{\bullet\bullet}{O}}\overset{\bullet}{\underset{\bullet}{\vdots}}$$

The observed bond distance is 0·114 nm. The bond length for the N=O bond would be expected to be 0·115 nm and for the N≡O bond, 0·105 nm.

(e) *The nitrate(V) ion.* A resonance hybrid between

$$O=N\underset{O^-}{\overset{O}{\diagup}} \longleftrightarrow {}^-O-N\underset{O}{\overset{O}{\diagup}} \longleftrightarrow O \leftarrow N\underset{O}{\overset{O^-}{\diagup}}$$

The observed bond length is 0·121 nm. The calculated bond length for N—O is 0·136 nm and for N=O, 0·115 nm. The resonance energy is 188·3 kJ mol^{-1}.

(f) *The nitro group.* A resonance hybrid between

$$-N\begin{matrix}\nearrow O\\ \searrow O\end{matrix} \longleftrightarrow -N\begin{matrix}\nearrow O\\ \searrow O\end{matrix}$$

The N—O bond lengths in a variety of nitro-compounds vary between 0·121 and 0·123 nm. A single bond between nitrogen and oxygen atoms would be expected to give a bond length of 0·136 nm; a double bond would give 0·115 nm.

1, 4-dinitrobenzene has no dipole moment, which indicates that the dipole moments in the two nitro groups must be equal in magnitude, but opposite in direction. The dipole moment of each nitro group must, in fact, act in a direction which bisects the ONO angle. The dipole moments of each of the individual structures given above would act mainly along the direction of the dative bonds.

(g) *The carbonate ion.* A resonance hybrid between

$$O=C\begin{matrix}\nearrow O^-\\ \searrow O^-\end{matrix} \longleftrightarrow {}^-O-C\begin{matrix}\nearrow O\\ \searrow O^-\end{matrix} \longleftrightarrow {}^-O-C\begin{matrix}\nearrow O^-\\ \searrow O\end{matrix}$$

The calculated bond length for the C=O bond is 0·122 nm and for the C—O bond, 0·143 nm. The observed bond length in the carbonate ion is 0·131 nm. The resonance energy is 176 kJ mol^{-1}.

(h) *The sulphate(VI) ion.* In the sulphate(VI) ion, the four oxygen atoms are arranged around the central sulphur atom almost tetrahedrally. The measured bond distances are all equal to 0·151 nm. The calculated bond distance for single bonds between sulphur and oxygen atoms is 0·170 nm, and for double bonds, 0·149 nm.

It is clear that there must be some double-bond character about the bonds in the ion, and it is best represented as a resonance hybrid between many various structures of which some typical ones are shown:

$$\left\{O\leftarrow\overset{\overset{\uparrow}{O}}{\underset{\underset{\downarrow}{O}}{S}}\rightarrow O\right\}^{2-} \quad \left\{O-\overset{\overset{\|}{O}}{\underset{\underset{|}{O}}{S}}\rightarrow O\right\}^{2-} \quad \left\{O-\overset{\overset{\|}{O}}{\underset{\underset{|}{O}}{S}}=O\right\}^{2-}$$

The last of these structures is probably the most important, but there are many similar ones because the double bonds may have alternative positions.

(i) *Benzene.* The properties of benzene can be accounted for, reasonably satisfactorily, by regarding it as a resonance hybrid between the two well-known Kekulé structures with three possible Dewar structures making a small contribution.

I II
Kekulé structures

One of three
Dewar structures

The bond length of a C—C bond would be expected to be 0·154 nm, and that of a C=C bond, 0·134 nm. The actual bonds between carbon atoms in benzene are all alike, with a bond length of 0·139 nm.

The actual measured heat of formation of benzene is 5501 kJ mol^{-1}, but the calculated heat of formation for structures I or II is only 5339 kJ mol^{-1}. The resonance energy of benzene is, therefore, 162 kJ mol^{-1}. A similar figure is obtained, too, from heats of hydrogenation. The expected heat of hydrogenation for structures I or II would be 372·9 kJ mol^{-1}, i.e. three times the heat of hydrogenation of a C=C bond, given on page 139 as 124·3 kJ mol^{-1}. The measured heat of hydrogenation of benzene is, however, only 208·4 kJ mol^{-1}. The resonance energy, from these figures, is 164·5 kJ mol^{-1}.

Representation of benzene as a resonance hybrid does not postulate the existence of true C=C bonds in a benzene molecule, and this accounts for the lack of true unsaturation properties such as those shown by ethylene.

IONIC CHARACTER OF COVALENT BONDS

5 Introduction A covalent bond between an atom A and an atom B involves sharing of electrons and is represented as $A \vdots B$. An ionic bond involves complete transfer of electrons and is represented as $[A]^+[\vdots B]^-$.

In a covalent bond between unlike atoms the pair of shared electrons will not necessarily be shared equally by both atoms, for if, in a bond A—B, the atom B has a stronger attraction for electrons than A the shared pair will be attracted towards B and away from A. Any permanent displacement of electrons of this sort in a covalent bond will give the bond some ionic character, and the actual bond will have to be represented as a resonance hybrid between covalent and ionic forms,

$$A\text{—}B \longleftrightarrow A^+B^-$$

In an alternative method of describing the same state of affairs an *inductive effect* is said to exist in the bond, and this is symbolised as

$$A{\rightarrow}B$$

In terms of molecular orbitals, the ionic character of a covalent bond can be represented by the shape of the orbital as indicated in Fig. 82.

Covalent
$A-B$

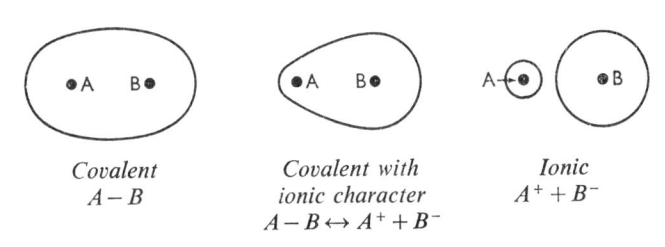

Covalent with
ionic character
$A-B \leftrightarrow A^+ + B^-$

Ionic
$A^+ + B^-$

Fig. 82. General shape of orbitals representing transition from pure covalent to pure ionic bonding

6 Electronegativity scales The extent of the ionic character, or the inductive effect, in a covalent bond will depend on the relative attraction for electrons of the bonded atoms. A covalent bond between like atoms will have no ionic character or inductive effect; it might be regarded as a true covalent bond. But in a bond between an electropositive atom, A, and an electronegative atom, B, the ionic structure A^+B^- may have a similar stability to the covalent structure A—B. If so, the ionic character of the actual bond will be high.

The stability of the A^+B^- form will clearly depend on the relative 'affinities' of A and B for electrons. Various terms are used to describe this, and atoms with a strong 'affinity' for electrons are said to be strongly electrophilic, to be electronegative, or to have a high electronegativity. The greater the difference between the electronegativities of the atoms A and B, the stabler A^+B^- (or A^-B^+) will be and the more ionic in character will the bond between A and B be.

In terms of wave functions, the value of λ in the equation

$$\psi = \psi_A \pm \lambda . \psi_B$$

is only 1 when A and B are alike. In other cases, the value of λ gives a measure of the ionic character of the bond between A and B. The value of λ cannot, unfortunately, be calculated from first principles, but it is possible to give quantitative values to the electronegativities of various atoms, and to draw up electronegativity scales. Two main methods have been adopted.

(a) *Pauling's method.* Pauling based his electronegativity scale on the estimated contribution of the A^+B^- structure to the actual bond existing between A and B. This is done by making the following measurements:

Actual bond energy measured experimentally $= H$
Bond energy if the bond was truly covalent $= Q$
Resonance energy caused by ionic character of bond $= H - Q$

Values of H can be measured experimentally, but a truly covalent bond between unlike atoms is hypothetical, and therefore the value of Q can only be obtained indirectly. Pauling has obtained values of Q by taking the bond energy of the truly covalent bond A—B (E_{AB}) as either the arithmetic or the geometric mean of the bond energies of the bonds A—A (E_{AA}) and B—B (E_{BB}), i.e.

$$E_{AB} = \frac{E_{AA} + E_{BB}}{2} \quad \text{or} \quad E_{AB} = \sqrt{(E_{AA} \times E_{BB})}$$

The difference between the values of H and Q, obtained in these ways, is then taken as the ionic resonance energy of the bond A—B. Typical values obtained are 267·8 kJ mol^{-1} for H—F and 6·7 kJ mol^{-1} for H—I. The smaller value for HI is to be expected, for iodine is well known to be less electronegative than fluorine, and the higher ionic resonance energies would be expected the greater is the difference between the electronegatives of the two atoms linked together.

Pauling has used the ionic resonance energies as a means of estimating this difference in electronegativity and has drawn up an electronegativity scale, given below, on the basis that the electronegativity, x, of an element is such that $(x_B - x_A)$ is approximately equal to the square root of the ionic resonance energy of the bond A—B. The ionic resonance energies used are expressed in electron-volts (pages 14).

(b) *Mulliken's method.* Mulliken defines the electronegativity of an atom as the arithmetic mean of its ionisation energy and its electron affinity (pages 44–8).

The ionisation energy, I_A, of an atom, A, is the energy required to convert the atom into an ion, A^+, whilst the electron affinity, E_B, of an atom B is the energy change in the reaction in which B^- is formed from B. Thus

$$A + \text{Ionisation energy } (I_A) \rightarrow A^+ + 1e$$
$$B + 1e \rightarrow B^- + \text{Electron affinity } (E_B)$$

To convert both A and B into A^+ and B^- requires, therefore, a total energy of $I_A - E_B$, whereas to obtain A^- and B^+ requires $I_B - E_A$. Mulliken argued that A and B would have equal electronegativities when $I_A - E_B$ was equal to $I_B - E_A$, i.e. when the same energy was required to form A^+B^- as to form A^-B^+. Thus, for equal electronegativities,

$$I_A - E_B = I_B - E_A \quad \text{or} \quad I_A + E_A = I_B + E_B$$

The greater the value of $I_A + E_A$, the greater the chance that it will be bigger than $I_B + E_B$, and if $I_A + E_A$ is bigger than $I_B + E_B$ then the easier it is to form A^-B^+. The value of $I_A + E_A$ gives, therefore, a measure of the electronegativity of the atom A.

The application of Mulliken's method is limited by the fact that it is difficult to measure values of electron affinities.

(c) *Electronegativity values.* The numerical values for electronegativities obtained by Pauling's and Mulliken's methods agree reasonably well though Mulliken's values have to be divided by 3·15 to give Pauling's values. Typical values based on Pauling's method are summarised below:

Electronegativity values (Pauling)

H						
2·1						

Li	Be	B	C	N	O	F
1·0	1·5	2·0	2·5	3·0	3·5	4·0

Na	Mg	Al	Si	P	S	Cl
0·9	1·2	1·5	1·8	2·1	2·5	3·0

K	Ca			As	Se	Br
0·8	1·0			2·0	2·4	2·8

Rb	Sr			Sb	Te	I
0·8	1·0			1·9	2·1	2·5

The greater the difference, $(x_B - x_A)$, between the electronegativity values of two atoms, A and B, the greater the ionic character of the bond between A and B. A number of numerical relationships between percentage ionic character and electronegativity have been suggested. According to one, an $(x_B - x_A)$ value of 1·7 leads to 50 per cent ionic character; a value of 2·3 to 73 per cent ionic character. On this basis the percentage of ionic character in typical bonds is given below

C—H	N—H	O—H	F—H
4%	19%	39%	60%

C—F	C—Cl	C—Br	C—I
43%	11%	3%	0%

7 Dipole moments A magnet has a magnetic moment equal to ml where m is the pole strength of the magnet and l the distance between the poles. In a similar way, two equal and opposite, but separated, electrical charges constitute an electrical dipole moment measured by the charge multiplied by the distance between the two charges.

Dipole moments are expressed in terms of coulomb metre (C m) units or the older debye unit, 1 D being equal to $3·335\,640 \times 10^{-30}$ C m. All dipole moments are of the order 10^{-30} C m, and this is to be expected for the charge on an electron is of the order 10^{-19} C and molecular distances are of the order 10^{-10} m.

Any bond which has any degree of polarity will have a corresponding dipole moment, though it does not follow that compounds containing such bonds will have dipole moments, for the polarity of the molecule as a whole is the vector sum of the individual bond moments. The C—Cl bond, for instance, has definite polarity, C→Cl or $C^{\oplus}$—$C^{\ominus}$, and a definite dipole moment. Tetrachloromethane, however, has no dipole moment for the resultant of the four C—Cl moments is zero. By comparison, mono-, di- and tri-chloromethane all have dipole moments.

Some typical values of dipole moments are given below, in C m × 10^{30}:

HF	6·369	HBr	2·601	H_2O	6·137
HCl	3·436	HI	1·268	NH_3	5·002

Measurements of dipole moments can be used to throw light on many problems in structural chemistry. Typical uses are indicated on pages 119–22.

8 Dipole moments and ionic character The measured dipole moment of a substance can, in simple cases, be used to estimate the ionic character of a bond. The dipole moment of HCl, for example, is $3·436 \times 10^{-30}$ C m. If the bond in the molecule was fully ionic the expected dipole moment would be equal to the charge on the electron multiplied by the bond distance, i.e.

$$(1{\cdot}602 \times 10^{-19}) \times (1{\cdot}29 \times 10^{-10}) = 20{\cdot}67 \times 10^{-30} \text{ C m}$$

The ionic character of the HCl bond may, therefore, be taken as $(3{\cdot}436/20{\cdot}67) \times 100$ per cent, i.e. approximately 17 per cent. This result is in good agreement with that obtained from the electronegativity values of hydrogen and chlorine.

It is also of interest to note that the dipole moment of a bond A—B, in debye units, is often approximately equal to the difference in the electronegativities of A and B, $(x_B - x_A)$, though this relationship cannot be used for quantitative purposes.

9 Oxidation number or oxidation state It is not always possible to express the numerical valency of an element in a compound without ambiguity; it depends, indeed, on what is to be meant by numerical valency (page 4). In the NO_3^-, ion, for example (page 120), the nitrogen atom is linked to *three* oxygen atoms, but it forms *four* bonds by using *five* of its own electrons.

The concept of oxidation number or oxidation state was introduced to provide a means of expressing, numerically, the way in which an element was combined within a compound.

In ionic compounds, the oxidation number of the elements concerned is equal to the electrical charge on their ions in the compound. In covalent compounds, oxidation numbers are taken, on an arbitrary basis, as being equal to the charges which the various atoms in a compound would carry if all the bonds in the compound were regarded as ionic instead of covalent. In doing this, a shared pair of electrons between two atoms is assigned to the atom with the greater electronegativity (page 125). Or, if the two atoms are alike, the shared pair is split between the two, one electron being assigned to each atom. The resulting charges on the various atoms when the bonding electrons are assigned in this way are the oxidation numbers of the atoms.

Some examples of oxidation numbers of atoms in different compounds will illustrate the usage:

C H$_4$	N H$_3$	H$_2$ O	H F
-4 $+1$	-3 $+1$	$+1$ -2	$+1$ -1
	Cl F	I Cl$_3$	I F$_5$
	$+1$ -1	$+3$ -1	$+5$ -1
	C H$_3$ Cl	C H$_2$ Cl$_2$	C Cl$_4$
	-2 $+1$ -1	0 $+1$ -1	$+4$ -1

The following points should be noted:

(a) The algebraic sum of the oxidation numbers of all the atoms in an uncharged compound is zero. In an ion, the algebraic sum is equal to the charge on the ion, e.g.

$$\underset{-3\ +1}{N\ H_4^+} \qquad \underset{-2\ +1}{O\ H^-}$$

(b) All elements in the elementary state have oxidation numbers of zero, shared pairs between like atoms being split equally between them.

(c) As fluorine is the most electronegative element it always has an oxidation number of -1 in any of its compounds.

(d) Oxygen, second only to fluorine in electronegativity, has an oxidation number of -2 in almost all its compounds,

$$\underset{+2\ -2}{Mg\ O} \quad \underset{+3\ -2}{Fe_2\ O_3} \quad \underset{+4\ -2}{C\ O_2} \quad \underset{+7\ -2}{Mn_2\ O_7} \quad \underset{+6\ -2}{Cr\ O_3}$$

Exceptions are provided by fluorine monoxide and the peroxides,

$$\underset{-1\ +2}{F_2\ O} \qquad \left[\underset{-1\quad -1}{:\!\overset{..}{O}\!:\!\overset{\times\times}{\underset{\circ\times}{O}}\!\overset{\times}{\underset{\times}{}}} \right]^{2-}$$

the oxidation number of the atoms in the peroxide ion being calculated by splitting the shared pair between the two oxygen atoms linked together.

(e) In all compounds, except ionic metallic hydrides, the oxidation number of hydrogen is $+1$, e.g.

$$\underset{+1\ -1}{H\ Cl} \quad \underset{+1\ -2}{H_2\ O} \quad \underset{-3\ +1}{N\ H_3} \quad \underset{+1\ -1}{Li\ H} \quad \underset{+2\ -1}{Ca\ H_2}$$

(f) In compounds containing more than two elements, the oxidation number of any one of them may have to be obtained by first assigning reasonable oxidation numbers to the other elements. In sulphuric(VI) acid, H_2SO_4, for example, the most reasonable oxidation numbers for hydrogen and oxygen are $+1$ and -2, which gives sulphur an oxidation number of $+6$. Other examples are,

$$\underset{+1\ +4\ -2}{H_2\ S\ O_3} \quad \underset{+1\ +7\ -2}{K\ Mn\ O_4} \quad \underset{+1\ +6\ -2}{K_2\ Cr_2\ O_7} \quad \underset{+1\ +7\ -2}{K\ Cl\ O_4}$$

(g) Some elements may have widely different oxidation numbers in different compounds as is shown by the following compounds of manganese, chromium, nitrogen and chlorine:

−3	−2	−1	0	1	2	3	4	5	6	7
			Mn		MnCl$_2$	MnCl$_3$ Mn$_2$O$_3$	MnO$_2$		MnO$_4^{2-}$	MnO$_4^-$ M$_2$O$_7$
			Cr		CrCl$_2$	CrCl$_3$ Cr$_2$O$_3$			CrO$_3$ CrO$_4^{2-}$ Cr$_2$O$_7$	
NH$_3$ NH$_4^+$	N$_2$H$_4$	NH$_2$OH	N$_2$	N$_2$O	NO	N$_2$O$_3$ HNO$_2$	NO$_2$	N$_2$O$_5$ HNO$_3$		
		HCl	Cl$_2$	HClO		HClO$_2$	ClO$_2$	HClO$_3$		HClO$_4$

(*h*) *When an element is oxidised its oxidation number is increased. When an element is reduced its oxidation number is decreased.*

Change in oxidation number can be used to decide whether an oxidation or a reduction has taken place. In the change from chloromethane to dichloromethane, for example,

$$\begin{array}{ccc} C\ H_3\ Cl & \to & C\ H_2\ Cl_2 \\ -2\ +1\ -1 & & 0\ +1\ -1 \end{array}$$

the oxidation number of carbon is increased from −2 to 0. The carbon is therefore being oxidised.

10 The stock system of nomenclature Oxidation numbers provide the basis for the Stock system of naming chemical compounds, which is important, though not, as yet, universally adopted. Some examples of Stock names will illustrate the system.

(*a*) *Binary compounds of metals with non-metals:*

MnCl$_2$ Manganese (II) chloride   Fe$_2$O$_3$ Iron (III) oxide
MnCl$_3$ Manganese (III) chloride   Fe$_3$O$_4$ Iron (II, III) oxide
MnO$_2$ Manganese (IV) oxide   PbO$_2$ Lead (IV) oxide
Mn$_2$O$_7$ Manganese (VII) oxide   Pb$_3$O$_4$ Dilead (II), lead (IV) oxide

(*b*) *Binary compounds of non-metals with non-metals:*

N$_2$O   Nitrogen (I) oxide   Cl$_2$O Chlorine (I) oxide
NO   Nitrogen (II) oxide   ClO$_2$ Chlorine (IV) oxide
NO$_2$ Nitrogen (IV) oxide

(*c*) *Anions and cations:*

NO$_2^-$ Nitrate (III) ion   SO$_3^{2-}$ Sulphate (IV) ion
NO$_3^-$ Nitrate (V) ion   SO$_4^{2-}$ Sulphate (VI) ion
Cu(H$_2$O)$_4^{2+}$ tetraquocopper (II) ion   Ag(NH$_3$)$_2^+$ diamminesilver (I) ion

Other examples of Stock names for complex ions are given on page 164.

12 Characteristics of Covalent Compounds

1 Molecular crystals When a substance is made up of a large number of individual, covalent molecules, the solid form of it is either amorphous or exists in what are known as molecular crystals. The atoms within each separate molecule are united by covalent bonds, but the individual molecules are held together, within the crystal, by weak intermolecular forces discussed in Chapter 14. Molecular crystals are formed by many organic substances and by many non-metallic elements. Their crystal shapes are discussed in Chapter 14.

Because the intermolecular forces in the crystal are very weak, substances which form molecular crystals have low melting and boiling points; little energy is needed to break down the crystal structure. Indeed, many of the substances concerned exist as gases or liquids at room temperature. If they are solid at room temperature, they are often soft.

The substances are non-electrolytes, and are commonly soluble in benzene and other organic solvents but insoluble in water and other ionising solvents. Covalent substances, e.g. hydrogen chloride may, however, appear to be soluble in water because they react with it.

A comparison of the properties of sodium chloride and tetrachloromethane shows, in a rather exaggerated way, the typical differences between an electrovalent compound forming ionic crystals and a covalent compound forming molecular crystals.

	Sodium chloride	Tetrachloromethane
(a)	Composed of Na^+ and Cl^- ions in an ionic crystal	Composed of individual CCl_4 molecules with weak intermolecular forces. Forms a molecular crystal when solid
(b)	Electrolyte	Non-electrolyte
(c)	Hard solid at room temperature	Liquid at room temperature. Soft, when solid
(d)	m.p. = 803°C b.p. = 1430°C	m.p. = −28°C b.p. = 77°C
(e)	Soluble in water. Insoluble in benzene	Soluble in benzene. Insoluble in water

2 Atomic or covalent crystals In substances which form molecular crystals, covalent bonds hold atoms together in discrete molecules. Covalent bonds can, however, hold atoms together in interlocking structures known as atomic or covalent crystals. In such a structure, e.g. the diamond structure (page 83), each atom is linked by covalent bonds to all its nearest neighbours.

No single molecules of such substances exist; they are sometimes said to be made up of *giant molecules*. Moreover, the covalent bonds holding the whole crystal together are strong so that such substances are hard solids, and have high melting and boiling points. Like all covalent substances, they are non-electrolytes.

The three main structures involved in atomic or covalent crystals are the diamond, zinc-blende and wurtzite structures.

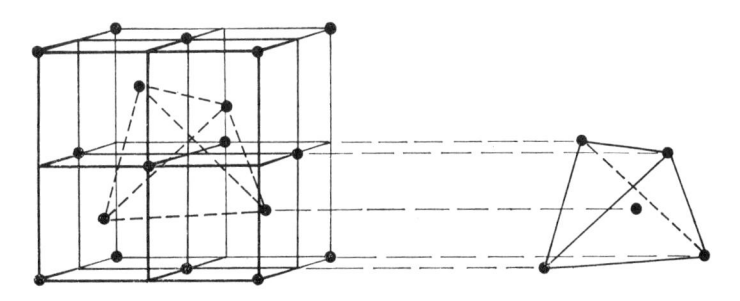

Fig. 83. The crystal structure of diamond showing (right) *the tetrahedral arrangement of four carbon atoms around a central carbon atom*

(a) *The diamond structure.* In a diamond crystal (Fig. 83) each carbon atom is surrounded by four other carbon atoms arranged tetrahedrally. The atoms are linked by covalent bonds, which are sp^3-hybrid bonds (page 112).

Germanium, silicon and grey-tin also have the diamond structure, though it is common for Group 4 elements to exhibit allotropy. Carbon atoms, for example, can be linked together in a different structure, as in graphite (page 157).

(b) *The zinc-blende structure.* Zinc-blende is one form of zinc sulphide. The crystal structure (Fig. 84) is related to that of diamond, the only difference being that adjacent atoms are different. Thus each zinc atom is surrounded by four sulphur atoms arranged tetrahedrally, and each sulphur atom by four zinc atoms similarly arranged.

Other common substances which crystallise with a zinc-blende structure are copper(I) chloride, bromide and iodide, aluminium phosphide, silicon carbide, and the sulphides of beryllium, cadmium and mercury(II).

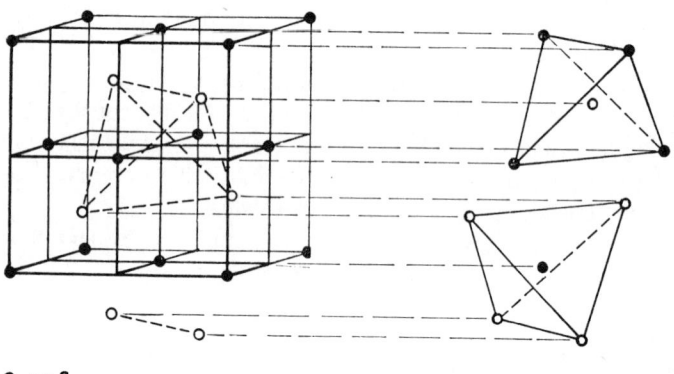

• = S
○ = Zn

Fig. 84. The crystal structure of zinc blende showing (upper right) *the tetrahedral arrangement of four sulphur atoms about a zinc atom and* (lower right) *the tetrahedral arrangement of four zinc atoms about a sulphur atom. Compare the crystal structure of diamond* (Fig. 83)

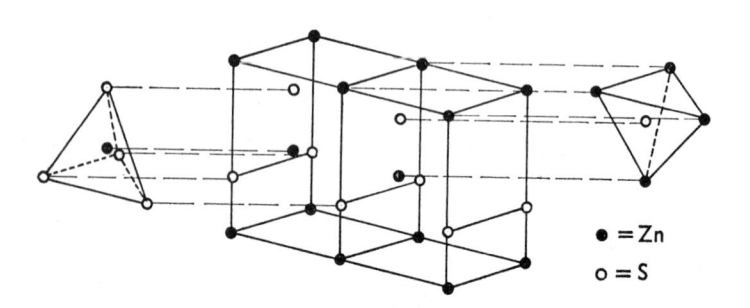

• = Zn
○ = S

Fig. 85. The crystal structure of wurtzite showing (right) *the tetrahedral arrangement of four zinc atoms around a sulphur atom, and* (left) *the tetrahedral arrangement of four sulphur atoms around a zinc atom*

(c) *The wurtzite structure.* Wurtzite is another form of zinc sulphide, and the structure (Fig. 85) is closely related to the zinc-blende structure (page 131). Each atom is surrounded by four nearest neighbours, arranged tetrahedrally, as in zinc-blende. It is only when the second nearest neighbours are considered that the wurtzite structure differs from that of zinc-blende.

Silver iodide, beryllium oxide, zinc oxide, cadmium sulphide and aluminium nitride are amongst the common compounds which form crystals with the wurtzite structure.

BOND LENGTHS

3 Single-bond covalent radii The bond length of a covalent bond is the distance between the nuclei of the two bonded atoms. The distances involved are very small and they are usually expressed in nanometres (page 7).

Bond lengths can be measured by X-ray analysis of crystals, by the diffraction of X-rays, electrons or neutrons by gases or vapours, or by spectroscopic methods. The results of measurements on a wide variety of compounds show that the inter-atomic distance between two atoms A and B linked by a single covalent bond is very nearly constant, and is independent of the varied nature of the molecules in which the bond length is measured. The C—Cl distance, for instance, is 0·176 nm in mono-, di-, tri- and tetrachloromethane. Similarly the C—C distance is 0·154 nm whether the bond occurs in diamond, propane, 1,3,5-trimethylbenzene, ethanal trimer, or many other compounds.

It is found, moreover, that the interatomic distance A—B is equal to the arithmetic mean of the A—A and B—B distances, i.e.

$$A{-}B = \frac{A{-}A + B{-}B}{2}$$

and, because of this simple relationship, it is possible to allot what are called covalent radii to the elements such that $r_A + r_B = A{-}B$. The covalent radii of those elements which normally form covalent bonds are summarised below.

	H 0·030	B 0·080	C 0·077	N 0·074	O 0·074	F 0·072
			Si 0·117	P 0·110	S 0·104	Cl 0·099
Single bond covalent radii (nm)				As 0·121	Se 0·117	Br 0·114
				Sb 0·141	Te 0·137	I 0·133

It must be emphasised that the values given only refer to elements when they exhibit their normal covalencies, i.e. when they form only single bonds equal in number to the group number in which the element is placed in the periodic table. They are referred to as single-bond covalent radii or normal covalent radii.

As is to be expected, the heavier elements in any one group of the periodic table have the larger radii; the heavier elements contain more extra-nuclear electrons. For elements in the same horizontal periods those with higher atomic numbers have the smaller radii; these elements have their outer electrons in the same orbits, but the electrons are more strongly attracted by increasing positive charges on the nuclei.

4 Double and triple bond radii Measurement of the interatomic distance between two atoms joined by double or triple bonds leads to results similar to those obtained for single bonds. A double or triple bond between any two atoms has a constant length, and the arithmetic mean relationship holds as for single bonds.

It is therefore possible to assign double-bond or triple-bond covalent radii to the elements and some numerical values are given below:

Multiple-Bond Covalent Radii
(Values in nm)

B=	C=	N=	O=
0·076	0·067	0·060	0·055
B≡	C≡	N≡	O≡
0·068	0·060	0·055	0·050
	Si=		S=
	0·107		0·094

By comparing the values given with those for single-bond covalent radii it will be seen that the double-bond radius of an atom is about 13 per cent and the triple-bond radius about 22 per cent less than the corresponding single-bond radius.

5 Effect of ionic character on normal covalent radii The sum of the normal covalent radii of two atoms A and B only gives the bond length, A—B, when the bond is purely covalent or nearly so. In other cases the measured bond length is less than the sum of the covalent radii for the bonded atoms, an effect which is attributed to the partial ionic character of the bond (see page 122).

Schomaker and Stevenson have attempted to treat the effect quantitatively, and express the length of a bond A—B as

$$A-B = r_A + r_B - 0.09(x_A - x_B)$$

where r_A and r_B are the covalent radii of A and B, and x_A and x_B their respective electronegativities (see page 125). This expression, which is purely empirical, gives good agreement between calculated and measured bond lengths in many cases, but it fails to account for all the measured bond lengths.

BOND ENERGIES

6 Meaning of terms The bond energy of a covalent bond, A—B, in in a diatomic molecule is equal to E_{AB} where,

$$\underset{\text{(atom)}}{A} + \underset{\text{(atom)}}{B} \rightarrow A-B + E_{AB}$$

It is generally expressed in kJ mol^{-1}.

In a polyatomic molecule, containing more than one covalent bond, the term bond energy can be interpreted in two ways. The C—H bond energy in methane, for example, might be taken as one quarter of the heat of formation of methane from its atoms, i.e. one quarter of R in the equation,

$$\underset{\text{(atom)}}{C} + \underset{\text{(atoms)}}{4H} \rightarrow CH_4 + R$$

Such a bond energy is really an *average bond energy*. Its value for C—H is 413·4 kJ mol^{-1} (page 137).

Alternatively, the C—H bond energy in methane might be taken as the energy required to split one C—H bond in the molecule to form CH_3 and H, i.e. as the heat of reaction D_{CH_3-H},

$$CH_4 \rightarrow CH_3 + H - D_{CH_3-H}$$

This value is correctly called the *bond-dissociation energy*; its value for the C—H bond in methane is 431 kJ mol^{-1} (page 140).

The bond energy and the bond-dissociation energy will be alike for the bond in a diatomic molecule, but it will be coincidental if they are alike for the bonds in polyatomic molecules. Both values are useful. In what follows, however, average bond energies are mainly concerned. Some further details of bond-dissociation energies are given on page 140.

7 Measurement of bond energy in diatomic molecules Determination of the bond energy in a diatomic molecule containing a covalent bond involves the measurement of the heat of formation of the molecule from its free atoms, or, alternatively, the heat of dissociation of the molecule into its free atoms. The bond energy A—B is equal to H where

$$A + B \rightarrow A\text{—}B + H$$
$$\text{(atom)} \quad \text{(atom)} \quad \text{(molecule)}$$

Normally this cannot be measured directly, as a compound is formed from molecules and not from free atoms, and also because a compound splits up on dissociation into molecules, and not free atoms, of its component elements.

To measure the heat of the reaction

$$H_2 + Cl_2 \rightarrow 2HCl + H$$

for instance, does not give the bond energy of the H—Cl bond. What is required is the heat of the reaction

$$H + Cl \rightarrow HCl + Q$$

and to obtain this it is necessary to measure the heats of atomisation of both hydrogen and chlorine,

$$H_2 \rightarrow H + H - S$$
$$Cl_2 \rightarrow Cl + Cl - T$$

Q then becomes equal to $\frac{1}{2}(H+S+T)$.

The measurement of heats of atomisation is not easy, as a molecule cannot be completely dissociated into its free atoms merely by heating. The values can be obtained, however, from the change in the degree of dissociation with temperature, from the heat of sublimation, or by spectroscopic methods. Once the heats of atomisation have been obtained it is comparatively easy to measure the necessary heat of reaction.

8 Determination of the average bond energy in a polyatomic molecule

Average bond energies can be obtained from measured heats of formation and heats of atomisation. To obtain the value for the C—H bond in methane requires measurement of the heats of combustion of methane, solid carbon (graphite) and gaseous hydrogen, together with the heats of atomisation of graphite and gaseous hydrogen, as summarised below:

(a) $CH_4 + 2O_2 \rightarrow CO_2 + 2H_2O + 891$ kJ

(b) $\underset{\text{(solid)}}{C} + O_2 \rightarrow CO_2 + 395$ kJ

(c) $2H_2 + O_2 \rightarrow 2H_2O + 571.6$ kJ

(d) $2H_2 \rightarrow 4H - 864.4$ kJ $\quad \begin{cases} \text{Heat of atomisation of gaseous} \\ \text{hydrogen} = 432.2 \text{ kJ} \end{cases}$
(gas)　(atoms)

(e) $C \rightarrow C - 713.6$ kJ $\quad \begin{cases} \text{Heat of atomisation of solid} \\ \text{carbon} = 713.6 \text{ kJ} \end{cases}$
(solid)　(atom)

$(b) + (c) - (a)$ gives the heat of formation of methane from solid carbon and gaseous hydrogen, i.e.

(f) $C + 2H_2 \rightarrow CH_4 + (395 + 571.6 - 891)$ kJ, i.e. 75.6 kJ
(solid) (gas)

$(f) - (d) - (e)$ gives the heat of formation of methane from carbon atoms and hydrogen atoms, i.e.

$C + 4H \rightarrow CH_4 + (75.6 + 864.4 + 713.6)$ kJ, i.e. 1653.6 kJ
(atom)　(atoms)

One quarter of this gives the average C—H bond energy of 413.4 kJ mol^{-1}.

9 Values of bond energies Typical bond energies obtained either from measurements on diatomic molecules or by taking average bond energies from measurements on polyatomic molecules are summarised below:

H—H	C—C	N—N	O—O	S—S	F—F	F—H
432.2	341.4	160.7	146.0	266.9	139.3	554.0
	C—N	N—H	O—H	S—H	Cl—Cl	Cl—H
	290.0	388.7	461.2	366.1	241.8	432.6
	C—O				Br—Br	Br—H
	341.0				192.9	365.3
	C—H	Si—Si			I—I	I—H
	413.4	177.8			151.5	298.7

BOND ENERGIES OF SINGLE BONDS. (Values in kJ mol^{-1}.)

Similar methods can also be adopted to give values for multiple bonds:

C=C	C=N	N=N	C=O	O=O
611.3	564.8	408.4	723.8	401.7
	C≡C	C≡O	N≡N	
	803.7	1071	941.4	

BOND ENERGIES OF MULTIPLE BONDS. (Values in kJ mol^{-1}.)

10 Use of bond energies All the values given for bond energies in the preceding section are from data obtained by measurements on compounds for which it is possible to allot a definite structure. The particular use of the values obtained is that they are additive for a wide range of compounds. Heats of formation for such compounds can be calculated and the results obtained agree remarkably well with experimental measurements.

Ethanol will be taken as an illustrative example. Its heat of combustion is 1410 kJ mol^{-1}, i.e.

$$C_2H_5OH + 3O_2 = 2CO_2 + 3H_2O \text{ (liquid)} + 1410 \text{ kJ} \quad \text{(I)}$$

The heats of combustion of solid carbon and gaseous hydrogen are 395 and 286·1 kJ mol^{-1} respectively, i.e.

$$C \text{ (solid)} + O_2 = CO_2 + 395 \text{ kJ}$$

and

$$H_2 \text{ (gaseous)} + \tfrac{1}{2}O_2 = H_2O \text{ (liquid)} + 286\cdot1 \text{ kJ}.$$

Or, doubling the first and trebling the second equation,

$$2C \text{ (solid)} + 2O_2 = 2CO_2 + 790 \text{ kJ} \quad \text{(II)}$$

and

$$3H_2 \text{ (gaseous)} + 1\tfrac{1}{2}O_2 = 3H_2O \text{ (liquid)} + 858\cdot3 \text{ kJ} \quad \text{(III)}$$

Thus, by combining (I), (II), and (III),

$$2C \text{ (solid)} + 3H_2 \text{ (gaseous)} + \tfrac{1}{2}O_2 = C_2H_5OH + 238\cdot3 \text{ kJ} \quad \text{(IV)}$$

238·3 kJ mol^{-1} is, therefore, the heat of formation of ethanol from its elements in their normal form.

The heats of atomisation of solid carbon, gaseous hydrogen, and gaseous oxygen are 713·6, 216·1, and 247·3 kJ mol^{-1}, i.e.

$$2C \text{ (solid)} = 2C \text{ (atoms)} - 1427\cdot2 \text{ kJ}$$
$$3H_2 \text{ (gaseous)} = 6H \text{ (atoms)} - 1296\cdot6 \text{ kJ}$$
$$\tfrac{1}{2}O_2 \text{ (gaseous)} = O \text{ (atom)} - 247\cdot3 \text{ kJ}$$

Combining these three equations with (IV) gives the result

$$2C \text{ (atoms)} + 6H \text{ (atoms)} + O \text{ (atom)} = C_2H_5OH + 3209\cdot4 \text{ kJ}.$$

3209·4 kJ mol^{-1} is, therefore, the heat of formation of ethanol from its atoms.

The result calculated from bond energies is obtained simply by adding the energies of five C—H, one C—C, one C—O, and one O—H bonds, i.e. $(5 \times 413\cdot4) + 341\cdot4 + 341\cdot0 + 461\cdot2$. This gives the value of 3210·6 kJ mol^{-1} for the heat of formation of ethanol, which is in very good agreement with the previous value.

Heats of reaction can also be predicted from a knowledge of bond energies. The reduction of an ethylenic hydrocarbon to a paraffin, for instance, involves bond changes represented as follows:

$$\underbrace{\underset{611\cdot 3}{C=C} + \underset{432\cdot 6}{H-H}}_{1043\cdot 9} \rightarrow \underbrace{\underset{341\cdot 4}{C-C} + \underset{2\times 413\cdot 4}{2C-H}}_{1168\cdot 2}$$

It will be seen that the reduction involves an evolution of 124·3 kJ mol^{-1} and this is in very good agreement with experimental values.

11 The effect of resonance on the use of bond energy values There is, as shown in the preceding section, a remarkably good agreement between the measured and calculated values for the heat of formation of ethanol. In other compounds this may not be so.

The measured heat of formation of benzene, for example, is 5501 kJ mol^{-1} but the calculated value, on the basis of six C—H, three C—C and three C=C bonds, is only 5339 kJ mol^{-1}. Clearly the structure which has been used to calculate the theoretical value for the heat of formation cannot be the actual structure which occurs in benzene. The actual structure is, in fact, a resonance hybrid, with a resonance energy of 5501–5339, i.e. 162 kJ mol^{-1}.

A similar figure is obtained, too, from heats of hydrogenation. The expected heat of hydrogenation of a structure with three C=C bonds would be 372·9 kJ mol^{-1}, i.e. three times the heat of hydrogenation of a C=C bond, given above as 124·3 kJ mol^{-1}. The measured heat of hydrogenation of benzene is, however, only 208·4 kJ mol^{-1}, giving a resonance energy of 164·5 kJ mol^{-1}.

In general, when the experimentally measured values of the heats of formation or hydrogenation of a compound are greater than the heats of formation or hydrogenation calculated from the bond energies of the bonds in a structure purporting to represent the compound, then that single structure is probably only one of many contributing to a resonance hybrid.

12 The bond energy of hybrid bonds The same two atoms are not commonly found bonded together, in different compounds, by different types of hybrid bond, but the C—H bonds in ethyne, ethene and ethane do represent different bond types. In ethyne sp-hybrid orbitals are involved (page 109); in ethene, sp^2-hybrids (page 111); in ethane, sp^3 (page 112). The corresponding bond energies for the C—H bonds are

In ethyne (sp)   In ethene (sp^2)   In ethane (sp^3)
502 kJ mol^{-1} 443·5 kJ mol^{-1} 431 kJ mol^{-1}

In the CH radical, involving, presumably, unhybridised p orbitals, the bond energy, equal to the bond dissociation energy (page 135), is 338·9 kJ mol^{-1}.

The increase in bond energy in passing from p to sp^3 to sp^2 to sp-orbitals can be related to the increased overlap of such orbitals when involved in bond formation.

13 Bond-dissociation energies As explained on page 135, the bond energy of a bond is only equal to the bond-dissociation energy for a diatomic molecule. In polyatomic molecules an average bond energy might be taken, but different bond-dissociation energies will occur.

In methane, for example, four different bond-dissociation energies are involved, as summarised below:

$$CH_4 \rightarrow CH_3 + H \qquad D_{CH_3-H} = 431\cdot0 \text{ kJ}$$
$$CH_3 \rightarrow CH_2 + H \qquad D_{CH_2-H} = 364\cdot0 \text{ kJ}$$
$$CH_2 \rightarrow CH + H \qquad D_{CH-H} = 523\cdot0 \text{ kJ}$$
$$CH \rightarrow C + H \qquad D_{C-H} = 338\cdot9 \text{ kJ}$$

It will be seen that none of these bond-dissociation energy values is equal to the average bond energy for the C—H bond in methane (413·4 kJ mol^{-1}), but the sum of all four bond-dissociation energies (1656·9 kJ) is equal to $4 \times 413\cdot4$ (1653·6 kJ).

Similarly, the average bond energy for the O—H bond in water is 461·2 kJ mol^{-1}, whereas the two bond-dissociation energies concerned are

$$H_2O \rightarrow OH + H \qquad D_{HO-H} = 495\cdot9 \text{ kJ}$$
$$OH \rightarrow O + H \qquad D_{H-O} = 426\cdot5 \text{ kJ}$$

As would be expected, $2 \times 461\cdot2$ is equal to $495\cdot9 + 426\cdot5$.

13 Hydrogen Bonding

1 The valency of hydrogen Hydrogen is normally monovalent, forming only one bond, but it can do this in four different ways.

(*a*) *Formation of H^+ cations.* The H^+ ion is formed from a hydrogen atom by the loss of an electron and it is simply a proton (see page 5). In the presence of water it is invariably hydrated to form an $[H_3O]^+$ ion with a structure represented by

$$\left[H \leftarrow O \begin{matrix} H \\ \\ H \end{matrix} \right]^+$$

This ion is called the hydroxonium or hydronium ion and is normally what is meant when reference is made to a hydrogen ion. By hydration, the bare proton attains the electronic structure of helium.

The hydroxonium ion is present in solutions of all acids in water, and an acid can be defined as a substance which when dissolved in water produces this ion, e.g.

$$HCl + H_2O \rightarrow [H_3O]^+ + Cl^-$$
$$H_2SO_4 + 2H_2O \rightarrow 2[H_3O]^+ + SO_4^{2-}$$

(*b*) *Formation of H^- anions.* The negatively charged ion of hydrogen is formed from a hydrogen atom by the gain of one electron and it has the electronic structure of helium. It is present in a series of metallic hydrides known as the salt-like hydrides. Such hydrides are formed, as colourless crystals, by the alkali and alkaline earth metals simply by heating the metal in hydrogen:

| LiH | NaH | KH | RbH | CsH |
| CaH$_2$ | SrH$_2$ | BaH$_2$ | | |

The alkali metal hydrides have the same crystal structure as sodium chloride, the chloride ions, Cl^-, being replaced by the hydride ions, H^-. Further evidence of the existence of the H^- ion is provided by the fact that electrolysis of molten lithium hydride using steel electrodes produces lithium at the cathode and hydrogen at the anode. Lithium hydride must, therefore, be represented as an electrovalent compound, Li^+H^-, and similarly for the other hydrides.

The H^- ion cannot exist in aqueous solution for it reacts with water:

$$H^- + H_2O \rightarrow H_2 + OH^-$$

(c) *Formation of single covalent bond, H—*. The hydrogen atom has a single electron which can be shared with another electron, from some other atom, to form a single covalent bond, the hydrogen thereby attaining a noble gas (helium) structure. Hydrogen chloride, for example, has the structure

$$\text{H} : \overset{..}{\underset{..}{\text{Cl}}} :$$

and the hydrogen molecule also contains this type of bond, H:H.

This type of valency bond is to be found in the great majority of the hydrides of the elements known as the molecular or volatile hydrides, and including

B_2H_6	CH_4	NH_3	H_2O	HF
	SiH_4	PH_3	H_2S	HCl
	GeH_4	AsH_3	H_2Se	HBr
	SnH_4	SbH_3	H_2Te	HI
	PbH_4	BiH_3	(H_2Po)	

(d) *Formation of a one-electron bond, H·*. The existence of a valency bond made up of a single electron is very unusual. It undoubtedly occurs, however, in the hydrogen molecule-ion, H_2^+ described on page 91.

2 The hydrogen bond In the examples given in the preceding section hydrogen atoms form only one bond, but there are many well-known compounds in which hydrogen atoms appear to form two bonds.

Hydrogen fluoride, for example, is known, from relative molecular mass measurements, to be associated, i.e. to exist as $(HF)_n$, and the acid salt, potassium hydrogen fluoride, KHF_2, is also well known and must be derived from the acid H_2F_2.

Originally, $(HF)_n$ and KHF_2 were formulated by assuming that hydrogen could act as an acceptor and form a dative bond with fluorine acting as the donor:

$$(H\text{---}F\rightarrow H\text{---}F\rightarrow H\text{---}F)_n \qquad K^+(F\rightarrow H\text{---}F)^-$$

There is, however, no reason to assume that a hydrogen atom can act as an acceptor in this way for it has only one stable orbit ($1s$) which can hold only two electrons, and the formulae written above give hydrogen four electrons.

The linkage previously represented as a dative bond is now regarded as a special type of bond known as a hydrogen bond. The mechanism of its formation is thought to be mainly electrostatic. The hydrogen fluoride ion, for instance, is envisaged as two negatively charged flouride ions linked together by a positively charged hydrogen ion (proton). The proton is thought to be able to exert a sufficiently strong electrostatic attraction to do this because of its small size. Hydrogen bonding is, indeed, sometimes called proton bonding.

To distinguish a hydrogen bond it is best to write it as a dotted line so that the hydrogen flouride ion becomes $(F\cdots H-F)^-$, and, in general, when a hydrogen bond links two atoms, A and B, the structure is represented as $A-H\cdots B$ or as a resonance hybrid between $A-H\cdots B$ and $A\cdots H-B$.

That the electrostatic mechanism for the formation of a hydrogen bond is probably correct is shown by the fact that a hydrogen bond $A-H\cdots B$ is formed most easily if A and B have high electronegativities. Thus the tendency of an $A-H$ bond to form a hydrogen bond with another atom, B, increases rapidly from C—H through N—H and O—H to F—H, and it decreases in passing from O—H to S—H or from F—H to Cl—H. This shows that the bond $A-H$ has the greatest tendency to form hydrogen bonds when the ionic character of the bond is greatest, i.e. when the bond has the greatest polar character, $A^{\delta-}-H^{\delta+}$.

Fluorine, with the highest electronegativity, forms the strongest hydrogen bonds, and by far the greatest number of hydrogen bonds known are those which unite two oxygen atoms.

The hydrogen bond is a weak bond, the strength of the strongest being about 20–40 kJ mol^{-1} as compared with strengths of 150–500 kJ mol^{-1} for normal covalent bonds (page 137).

3 Inter-molecular hydrogen bonding. Association Examples of compounds in which a hydrogen bond is thought to be found between two, or more, molecules, with some of the evidence which points to the existence of such bonds are given below.

(*a*) *Hydrides of fluorine, oxygen and nitrogen.* The association of hydrogen fluoride has already been mentioned, and similar association is found in water and in ammonia. This shows itself in the high di-electric constants of these three hydrides and also in their abnormal melting and boiling points as compared with other hydrides in the same groups of the periodic table (Fig. 86).

The high melting and boiling points are due to association caused by the formation of hydrogen bonds, as shown, for water, in Fig. 87. Methane has normal values for its melting and boiling points. It is not associated, as carbon is not sufficiently electronegative to be linked by hydrogen bonds.

(*b*) *Ice.* The crystal structure of ice shows a tetrahedral arrangement of water molecules similar to that found in the wurtzite structure (page 132). Each oxygen atom is surrounded, tetrahedrally, by four

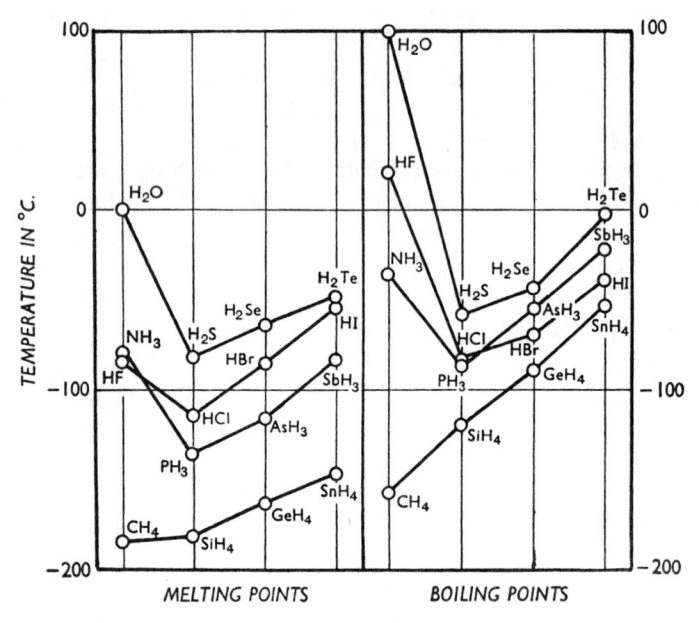

Fig. 86. The abnormal melting and boiling points of water, hydrogen fluoride and ammonia

others, and it is supposed that hydrogen bonds link pairs of oxygen atoms together as shown in Fig. 88.

The distance between adjacent oxygen atoms is 0·276 nm and this

Fig. 87. Hydrogen bonds causing association in water

suggests that the hydrogen atom linking the two oxygen atoms together is not midway between them, for the O—H distance in water vapour is 0·096 nm and not half 0·276 nm. Distance measurements on other compounds, too, indicate that the hydrogen atom in a hydrogen bonded pair of atoms is not equidistant from the two atoms.

The arrangement of the water molecules in ice is a very open structure and this explains the low density of ice. When ice melts, the structure breaks down and the molecules pack more closely together so that water has a higher density. This breaking down process is not complete until a temperature of 4°C is reached and it is on these lines that the abnormal behaviour of water is explained.

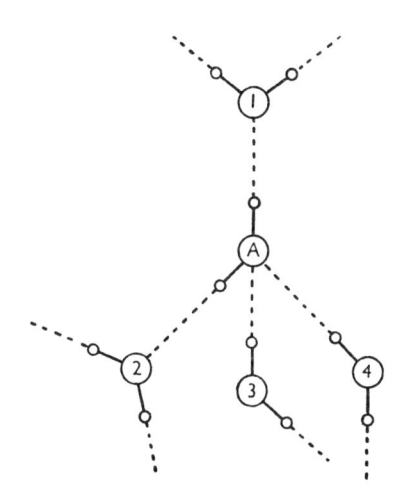

Fig. 88. The crystal structure of ice. The central oxygen atom, A, is surrounded tetrahedrally by the oxygen atoms, 1, 2, 3 and 4. All other oxygen atoms are arranged similarly. The hydrogen atoms are shown as small circles, and the dotted lines indicate the hydrogen bonds

(c) *Alcohols.* There is a marked difference between the boiling points of alcohols and the corresponding sulphur analogues (thiols), e.g.

	CH_3OH	C_2H_5OH	C_3H_7OH	C_4H_9OH
Boiling point	64·5°C	78°C	97°C	117°C

	CH_3SH	C_2H_5SH	C_3H_7SH	C_4H_9SH
Boiling point	5·8°C	37°C	67°C	97°C

Alcohols are, therefore, thought to be associated in much the same way as water.

(d) *Carboxylic acids.* Some carboxylic acids associate into dimers, i.e. two molecules link together, both in the vapour state and in certain solvents. Partition coefficient measurements of the distribution of ethanoic acid between water and benzene show, for instance, that the acid is present as a dimer in the organic solvent. This dimer is written as

$$H_3C-C\begin{array}{c}O-H\cdots O\\ \\O\cdots H-O\end{array}C-CH_3$$

and the presence of an eight-membered ring is confirmed by electron diffraction studies. Relative density measurements also show the presence of double molecules of ethanoic acid in the vapour state.

In aqueous solution, the molecules of a carboxylic acid link up with water molecules rather than form dimers.

(e) *Amines*. The association and basic strength of amines are both explained in terms of hydrogen bonding.

Primary and secondary amines are associated to some extent, though not greatly because nitrogen does not form hydrogen bonds very readily. Tertiary amines are not associated at all for they have no hydrogen atom capable of forming hydrogen bonds. Thus trimethylamine, which is not associated, has a lower boiling point (4°C) than dimethylamine (7°C) even though it has a higher relative molecular mass.

In aqueous solution, amines react with water molecules as shown,

$$CH_3.NH_2 + H_2O \rightleftharpoons CH_3.\overset{H}{\underset{H}{N}}-H\cdots O-H \rightleftharpoons (CH_3.NH_3)^+ + OH^-$$

The resulting solution will contain some $CH_3.NH_3OH$ molecules together with some $CH_3NH_3^+$ and OH^- ions. In a solution of a quaternary base, e.g. tetramethylammonium hydroxide, $[(CH_3)_4N]OH$, however, there are no hydrogen atoms which could form hydrogen bonds. As a result the solution contains only $(CH_3)_4N^+$ and OH^- ions. This explains why quaternary bases are very much stronger than primary, secondary or tertiary amines. It was, in fact, the marked basic strength of quaternary bases which first led Moore and Winmill (1912) to suggest the possible existence of hydrogen bonds.

(f) *Copper(II) sulphate(VI)-5-water*. Gentle heating of copper(II) sulphate(VI)-5-water will convert it into the monohydrate: it is, in fact, efflorescent in hot, dry climates. A much higher temperature is required, however, to remove the final molecule of water of crystallisation from the monohydrate. This suggests that one of the molecules of water of crystallisation is different from the other four, and leads to the writing of the formula as $Cu(H_2O)_4SO_4.H_2O$. The crystal structure is not very simple. Each Cu^{2+} ion is surrounded octahedrally by six oxygen atoms, four of them being in water molecules and two in sulphate ions. The extra water molecule is situated between two such octahedral groups. It is not attached directly to any Cu^{2+} ions; instead it is linked by hydrogen bonds between the water and sulphate groups surrounding the Cu^{2+} ions. $Cu(NH_3)_4SO_4.H_2O$ exists and has a structure like that of $Cu(H_2O)_4SO_4.H_2O$, but $CuSO_4.5NH_3$ does not exist. This is probably due to the fact that NH_3 molecules will not form hydrogen bonds so readily as H_2O molecules (see page 143).

Hydrogen bonding is also found in crystals of other salt hydrates, hydrated acids and hydroxides.

(g) *Ammonium fluoride.* Ammonium chloride, bromide and iodide have a caesium chloride structure at temperatures below 184·3°C, 137·8°C and −17·6°C respectively, but adopt a sodium chloride structure above these transition points. In ammonium fluoride the radius ratio is 0·92 so that a caesium chloride structure might be expected. Instead, ammonium fluoride crystallises with a wurtzite structure in which each NH_4^+ ion is surrounded tetrahedrally by four F^- ions. This difference is due to the formation of hydrogen bonds between the F^- ions and the hydrogen atoms of the NH_4^+ ion. Such bonds lock the NH_4^+ ions in position, but they are formed only in the fluoride because of the high electronegativity of fluorine.

4 Intra-molecular hydrogen bonding. Chelation Association, as described in the preceding section, occurs when hydrogen bonding takes place between two or more molecules, i.e. when there is intermolecular hydrogen bonding. Hydrogen bonding may, however, also take place within a single molecule; this is known as intra-molecular bonding. It may lead to the linking of two groups to form a ring structure. Such an effect is known as chelation, though the term is used in a wider sense (page 163) and this is but one kind of chelation. Examples are provided by the substances mentioned below.

(a) *1,2-substituted benzene compounds.* 2-nitrophenol boils at 214°C as compared with 290°C for 3- and 279°C for 4-nitrophenol. 2-nitrophenol is, moreover, volatile in steam, and less soluble in water than the other two isomers.

All these facts can be accounted for on the assumption that 2-nitrophenol contains an internal hydrogen bond represented as:

[Structure: 2-nitrophenol with intramolecular hydrogen bond between O–H and O=N→O]

This intra-molecular hydrogen bonding prevents inter-molecular bonding between two or more molecules. But in the 3- and 4-isomers, intramolecular bonding is not possible because of the size of the ring which would have to be formed. Inter-molecular bonding therefore takes place, and this causes some degree of association, which accounts for the higher boiling points of the 3- and 4-isomers.

The low solubility of 2-nitrophenol may be explained in two ways. The formation of an internal hydrogen bond 'suppresses' the hydroxylic character of the compound and this causes a lowering of solubility in water. In other words, the formation of an internal hydrogen bond prevents hydrogen bonding between 2-nitrophenol and water and this results in reduced solubility.

The effect of hydrogen bonding in 2-nitrophenol is also shown spectroscopically. A normal —OH group is found to give rise to a particular line in the infra-red absorption spectrum of the substance concerned. The spectra of 3- and 4-nitrophenol show this line, but that of 2-nitrophenol does not. There is not the same difference between the three methyl ethers of the nitrophenols because hydrogen bonding cannot take place in these ethers.

Other compounds in which intra-molecular hydrogen bonding plays the same part as in 2-nitrophenol include 2-hydroxy benzenecarbaldehyde (benzaldehyde), 2-chlorophenol and 2-hydroxy benzenecarboxylic acid (benzoic acid).

(*b*) *Ethyl 3-oxobutanoate*. Ethyl 3-oxobutanoate exists in two tautomeric forms known as the keto- and enol-forms:

$$CH_3.\underset{\underset{O}{\|}}{C}.CH_2.COOC_2H_5 \rightarrow CH_3.\underset{\underset{OH}{|}}{C}=CH.COOC_2H_5$$

(Ethyl 3-oxobutanoate)　　　(3-hydroxybut-2-enoate)

Meyer, in 1920, succeeded in separating these two forms by fractional distillation under reduced pressure in specially cleaned quartz apparatus (aseptic distillation).

Alcohols have, in general, higher boiling points than the corresponding ketones. Compare, for example, propan-2-ol (82°C) and propanone (56°C). But the 3-hydroxybut-2-enoate has a lower boiling point than ethyl 3-oxobutanoate. This is probably due to intra-molecular hydrogen bonding in 3-hydroxybut-2-enoate which prevents inter-molecular bonding and thus prevents association which would raise the boiling point. This intra-molecular hydrogen bonding is represented as:

$$H_3C-C\overset{\overset{\displaystyle H}{|}}{\underset{\underset{\displaystyle O-H\cdots O}{|}}{C}}\ \ \ \overset{}{\underset{\|}{C}}-OC_2H_5$$

The presence of such a bond is supported by the fact that 3-hydroxybut-2-enoate is less soluble in water, and more soluble in cyclohexane, than ethyl 3-oxobutanoate. This indicates a suppression of hydroxylic character in 3-hydroxybut-2-enoate.

5 Hydrogen bonding in proteins Protein molecules are made up from α-amino acid molecules, which are represented by the general formula:

$$R-\overset{\overset{\displaystyle H}{|}}{\underset{\underset{\displaystyle NH_2}{|}}{C}}-COOH$$

R may be a methyl, CH_3-, group, as in alanine, or a more complicated group, e.g. $HO.C_6H_4.CH_2-$ as in tyrosine, or $CH_3.S.CH_2.CH_2-$ as in methionine.

The $-NH_2$ group of one amino acid molecule can condense with the $-COOH$ group of another to eliminate water and form a peptide link

$$-\overset{O}{\underset{\|}{C}}-\overset{H}{\underset{|}{N}}-$$

between the two amino acids concerned. Two amino acids linked together by one peptide link make a dipeptide, but proteins are *polypeptides* with hundreds of peptide links between amino acid units. Any one protein may also contain 20 or so different amino acids, each acid contributing a number of different units to the long-chain molecule which may be represented as follows,

$$-\overset{O}{\underset{\|}{C}}-\overset{H}{\underset{|}{N}}-\overset{R}{\underset{|}{C}}-\overset{O}{\underset{\|}{C}}-\overset{H}{\underset{|}{N}}-\overset{R}{\underset{|}{C}}-\overset{O}{\underset{\|}{C}}-\overset{H}{\underset{|}{N}}-\overset{R}{\underset{|}{C}}-$$

One protein differs from another in the chain length and in the way in which the nature of the *R* groups varies along the chain. Cross links may occur between chains, particularly when the group *R* is sulphur-containing so that —S—S— bonds can form.

The linear sequence of amino acids, or *R* groups, along the chain of a protein is an important matter; it is known as the *primary structure* of the protein. The *secondary structure* refers to the detailed configuration of the chain, and one of the commonest configurations is a spiral form known as the α-helix form. The amino acid units are

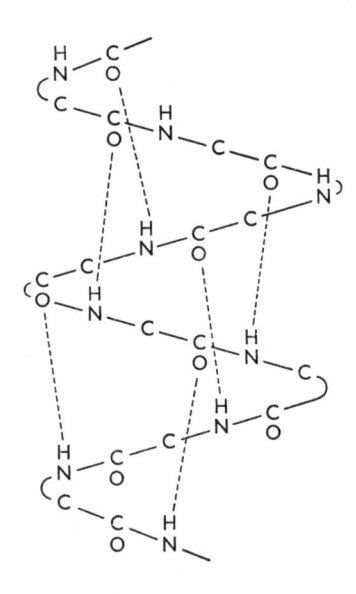

Fig. 89. Diagrammatic representation of the hydrogen bonding in the α-helix structure of a protein molecule

arranged around a helix, like a stretched coil spring. Such a molecular arrangement is stabilised by the existence of N—H····O hydrogen bonds between the N—H group of each amino acid unit and the fourth C=O group following it along the chain. The pattern repeats itself every five turns and there are 18 *R* groups within the five turns, or 3·6 *R* groups per turn (Fig. 89). The length of the repeating unit is 2·7 nm. The change from the helical arrangement to that of an extended chain, which only involves the breaking down of weak hydrogen bonds, probably accounts for the fact that many protein materials can stretch so freely.

The *tertiary structure* of a protein involves the way in which the helical or extended molecules are arranged in relation to each other. A number of helices can twist together like the wires in an electric cable. Extended molecules can pack together side by side in parallel or anti-parallel arrays known as pleated sheets. It is probable, too, that both the helical and extended molecules can twist into other shapes before packing together. In all cases, hydrogen bonding between adjacent molecules stabilises the structure. The secondary structure of a protein is, therefore, concerned with intra-molecular hydrogen bonding whilst the tertiary structure involves intermolecular bonding.

6 Hydrogen bonding in nucleic acids The inter-molecular bonding in the tertiary structure of proteins has a counterpart in the structure of nucleic acids. The molecule of deoxyribonucleic acid, DNA, for example, consists of two intertwining helices (a double helix) which are held together by hydrogen bonding.

Each helix is made up of nucleotides joined together in a particular sequence, and deoxythymidine-5′-phosphate(V) (Fig. 90) is a typical

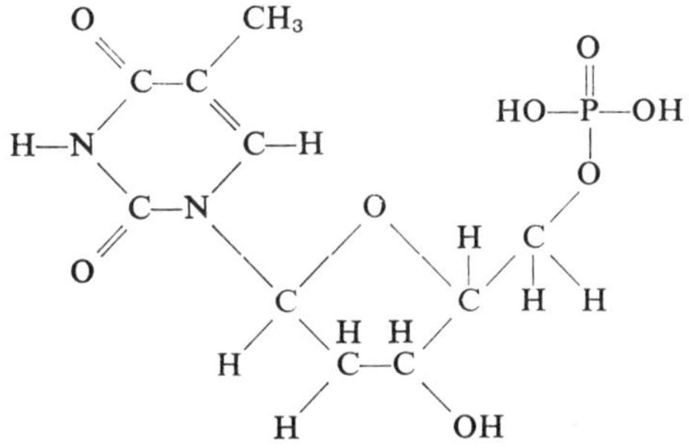

Fig. 90. Deoxythymidine 5′-phosphate(V). A typical nucleotide, which occurs in DNA. It is made up of the base, thymine (left), the sugar, deoxyribose (bottom) and phosphoric(V) acid (right)

nucleotide which occurs in DNA. Other nucleotides have identical phosphorylated sugar portions attached to different bases, such as adenine, guanine and cytosine, instead of thymine. The linking of one nucleotide to another is shown in Fig. 91.

In the two intertwining strands of the DNA molecule hydrogen bonding occurs between the basic parts of the nucleotides. But adenine can only form hydrogen bonds with thymine, and guanine with cytosine (base-pairing), as shown in Fig. 92, so that the sequences of base units in each strand must be complementary in this way; adenine must be opposite to thymine, and guanine to cytosine. A complete revolution of the molecular arrangement occurs about every ten nucleotide pairs (Fig. 93).

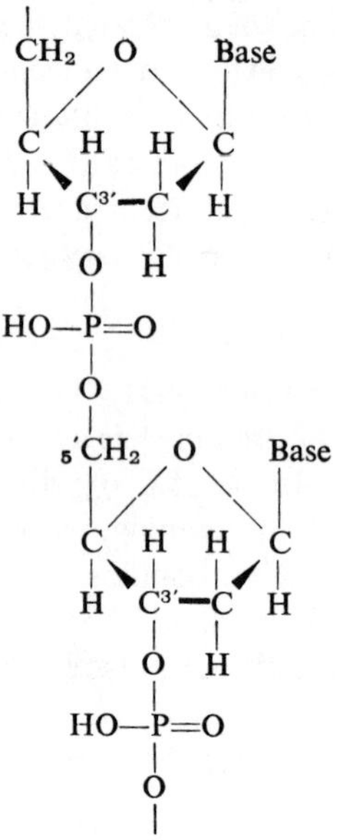

Fig. 91. The linking of one nucleotide to another in DNA

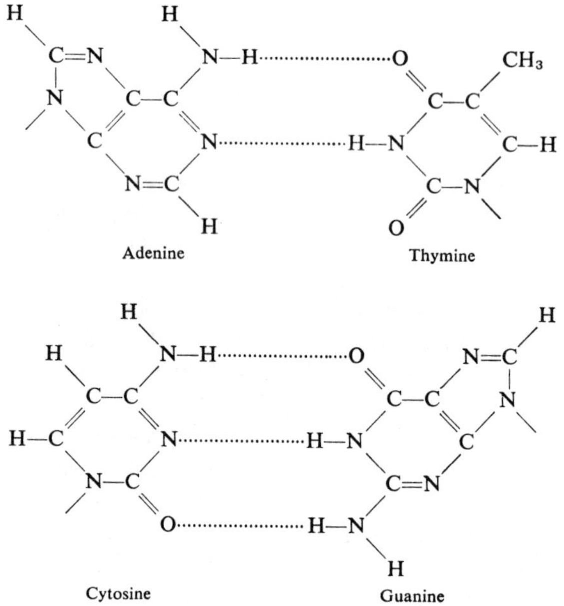

Fig. 92. Hydrogen bonding between thymine and adenine, and cytosine and guanine, in a DNA molucule

The breaking of the relatively weak hydrogen bonds in DNA plays a vital part in living processes, and this is probably but a pointer to the full role of the hydrogen bond in other biochemical changes still to be elucidated.

Fig. 93. The double helix of a DNA molecule. S *and* P *stand for sugar and phosphate respectively.* A, T, G *and* C *stand for the four bases which occur. The cross-linking between the two strands of the molecule provided by hydrogen bonding between pairs of bases is shown, the hydrogen bonds being given as dotted lines*

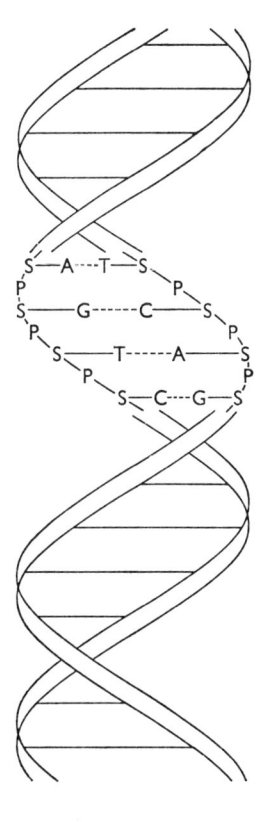

14 van der Waals' Forces

1 Introduction Substances such as oxygen, nitrogen, chlorine, iodine, and most organic compounds, e.g. naphthalene, exist in the gaseous state in the form of individual, covalently-bound molecules. The interaction between such molecules, in the gaseous state, is very small, but the fact that gases do not accurately obey the perfect gas laws is partially due to the existence of small cohesive forces between gas molecules. These forces are known as van der Waals' forces after the man who first took them into account in modifying the gas equation from $PV = RT$ to

$$\left(P + \frac{a}{V_2}\right)(V-b) = RT$$

When a gas or vapour is cooled, the molecular vibration decreases and the molecules come closer and closer together until they first condense into a liquid and then solidify. The van der Waals' forces are much stronger in the liquid and solid states than in the gaseous state. In the solid state, individual molecules are held together within a crystal by these van der Waals' forces; the crystals are known as *molecular crystals* (page 130).

2 Examples of molecular crystals The individual structural units in a molecular crystal may be atoms, as in solidfied noble gases, or

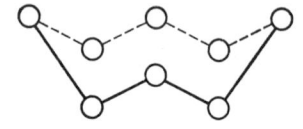

Fig. 94. S_8 molecule in a rhombic sulphur crystal

molecules. With atoms or small molecules, which can be regarded as approximately spherical, the crystal structure is an approximately close-packed arrangement (page 222), for the van der Waals' forces have no directional nature. Such structures are found in the solid forms of the noble gases, the halogens, hydrogen, nitrogen, oxygen, carbon monoxide, hydrogen chloride and bromide, methane, ethane, ammonia, phosphine and hydrogen sulphide.

With larger molecules as the structural units, their arrangement becomes more complicated, although still as close-packed as the shape of the molecules will allow. In rhombic sulphur, S_8 units occur

(Fig. 94); they break down into chains in plastic sulphur. In selenium and tellurium, the atoms are linked together in long puckered chains (Fig. 95), which extend right through the crystal. In white phosphorus, the structural unit is a P_4 molecule (Fig. 96). In organic solids, still

Fig. 95. Puckered chain of atoms which occur in crystals of Se *and* Te

more complex molecules occur, ranging from symmetrical molecules to long-chain, e.g. soap, molecules to flat, e.g. anthracene, molecules.

(a) *Noble gases.* The noble gases are monatomic and exist as single atoms in the gaseous state. In the solid state, these atoms are arranged in a cubic close-packed structure (page 223), each atom being surrounded by 12 equidistant neighbours.

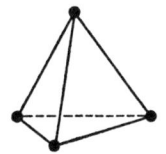

Fig. 96. P_4 *molecule in a white phosphorus crystal*

The relative weakness of the van der Waals' forces holding such a crystal together is shown by the low melting and boiling points of the noble gases; very little thermal energy is required to break down the crystal structure.

The melting and boiling points of the noble gases increase as the atomic number or size of the atom increases, as shown in the following table:

	He	Ne	Ar	Kr	Xe	Rn
Atomic no.	2	10	18	36	54	86
m.p./K	—	24·6	83·9	116	161	202
b.p./K	4·2	27·2	87·3	120	165	211

Such a variation suggests that the strength of the van der Waals' forces increases as the size of the particles linked together gets greater.

This is also seen by comparing the melting and boiling points of the halogens below and by the well-known rise in boiling and melting points in ascending a homologous series of organic compounds.

The van der Waals' forces between helium atoms are so weak that helium cannot be solidified at atmospheric pressure; at 26 atmospheres it solidifies at 0·9°K.

(b) *The halogens.* The actual crystal structures of the halogens are not all alike, but they all consist of diatomic molecules. Iodine is typical in having a close-packed structure very much like that of the inert gases. I_2 molecules pack together as shown in Fig. 97. The I—I

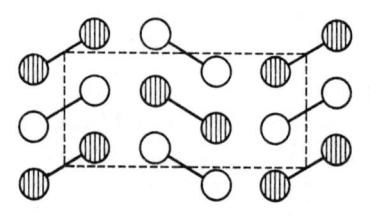

Fig. 97. The packing of I_2 molecules in an iodine crystal. The shaded molecules lie in the plane of the paper; the others lie below and above the plane

distance in the molecules in the crystal is 0·268 nm, which compares with a value of 0·266 nm for iodine molecules in the gaseous state.

As with the noble gases, the strength of the van der Waals' forces increases with atomic number, so that the melting and boiling points of the halogens increase in passing from fluorine to iodine:

	F	Cl	Br	I
m.p./°C	−218	−101	−7·3	114
b.p./°C	−188	−34·1	58·8	184

(c) *Graphite.* The building unit in the molecular crystals of graphite is an infinite layer of hexagonally arranged carbon atoms, linked within the layers by covalent bonds. The C—C distances within the layers are 0·142 nm; adjacent layers are held together, 0·335 nm apart, by van der Waals' forces (Fig. 98). The weakness of the forces allows the layers to slide over each other, which is why graphite is soft and acts as a solid lubricant.

The distance between the layers is also such that a number of graphitic 'compounds' can be formed in which other atoms or molecules take up a position between the layers. The layer structure is maintained, but there is some expansion in a direction perpendicular to the layers.

Graphite absorbs liquid potassium, for example, to form 'alloys' with detectable compositions represented by KC_8, KC_{16}, KC_{24} and

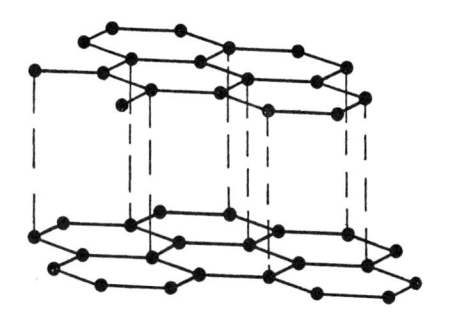

Fig. 98. Graphite

KC_{40}. Graphite also reacts with strong oxidising agents, such as nitric acid or potassium chlorate, to form graphitic oxides with compositions varying from $C_{2.9}O$ to $C_{3.5}O$. Similar 'compounds' are formed between graphite and fluorine, e.g. $(CF)_n$, and between graphite and acids such as sulphuric(VI) acid, e.g. $C_{24}HSO_4.2H_2SO_4$.

Boron nitride, BN, has a structure like that of graphite, boron and nitrogen atoms alternating within each of the layers.

(d) *Other layer lattices.* The type of structure found in graphite is sometimes known as a layer lattice structure, and other substances adopt the same form.

In cadmium chloride, $CdCl_2$, for example, Cd^{2+} ions are surrounded octahedrally by six Cl^- ions, the resulting $CdCl_6$ groups forming a layer as shown, two-dimensionally, in Fig. 99. Such layers

$$
\begin{array}{ccccc}
 & & Cl^- & & \\
 & Cl^- & & Cl^- & \\
Cd^{2+} & & Cd^{2+} & & Cd^{2+} \\
 & Cl^- & & Cl^- & \\
Cl^- & & Cl^- & & Cl^- \\
 & Cd^{2+} & & Cd^{2+} & \\
Cl^- & & Cl^- & & Cl^- \\
 & Cl^- & & Cl^- & \\
Cd^{2+} & & Cd^{2+} & & Cd^{2+} \\
 & Cl^- & & Cl^- & \\
 & & Cl^- & & \\
\end{array}
$$

Fig. 99. Two-dimensional representation of a single layer in the layer lattice of cadmium chloride. In three dimensions all the Cd^{2+} ions are in one plane, with the Cl^- ions in alternate horizontal rows above or below this plane

are held together in the crystal by van der Waals' forces, as in graphite. Similar arrangements occur in many di- and tri-chlorides, bromides and iodides, and in some sulphides and hydroxides. Molybdenum(IV) sulphide, MoS_2, provides a topical example. Because of its layer lattice it is used as a lubricant, particularly at high temperatures; it has recently been widely advertised as *Molyslip*.

In the hydroxides of zinc, beryllium, aluminium and iron(III), layers are held together by hydrogen bonds instead of van der Waals' forces.

(*e*) *Benzene.* The structural unit in benzene crystals is a hexagonal, planar arrangement of six CH groups; they pack together as shown in Fig. 100. Each circle represents a CH group, the plane of each C_6H_6

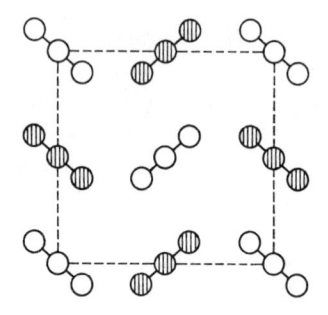

Fig. 100. Crystal structure of solid benzene. The planes of the C_6H_6 rings are perpendicular to the plane of the paper. The unshaded rings lie in the plane of the paper; the shaded ones are above and below

molecule being perpendicular to the plane of the paper. The shaded molecules are above and below the plane of the paper so that the central molecule is surrounded by 12 neighbours, four (unshaded) in the plane of the paper, four above and four below.

3 The nature of van der Waals' forces When the molecules held together in a molecular crystal are polar in nature, e.g. hydrogen chloride, water, ammonia, there will be a dipole–dipole attraction between the molecules when they are correctly orientated, and this provides a major contribution to the van der Waals' forces in such cases. The effect was first studied by Keesom in 1912; it is now referred to as the *orientation effect*.

The strength of the attractive forces between two polar molecules will also be increased by the fact that one dipole will induce another in a neighbouring molecule. Such effects, studied by Debye (1920), are known as *induction effects*.

With non-polar substances, such as iodine and the inert gases, there are no initial dipoles; the molecules are electrically symmetrical. Any slight relative displacement of the nuclei or the electrons in a molecule will, however, give rise to an electrical dipole, and it is thought that, momentarily, such displacements are constantly occurring within molecules or atoms. In a hydrogen atom, for example, consisting of one proton and one electron, the charge distribution of the electron may not be absolutely symmetrical at any one particular instant of time.

The electrical displacement necessary to establish a dipole is temporary and random, so that a molecule, as a whole, over a period of time, will have no observable dipole. But if it has a dipole at any one instant, it will induce another in a neighbouring molecule and an attractive force will be established between the two. Such interaction must be wholly responsible for van der Waals' bonding between non-polar atoms and molecules; its functioning was first explained by London (1930) and it is now referred to as the *dispersion effect*.

The dispersion effect makes a big contribution even to the bonding between polar molecules. Of the 29·7 kJ required to break down the crystal structure of solid ammonia, for example, it is estimated that 13·3 kJ are needed for the orientation effect, 1·55 kJ for the induction effect and 14·85 for the dispersion effect. The figure of 29·7 kJ gives a measure of the strength of the van der Waals' forces, and its comparison with the figure of 388·7 kJ for the bond energy of the N—H bond (page 137) shows how very much weaker van der Waals' forces are. For non-polar substances, the figures are even more marked. The van der Waals' bonding in solid oxygen, for example, corresponds to 7·1 kJ mol^{-1}; the O=O bond energy is 401·7 kJ mol^{-1} (page 137).

4 van der Waals' radii Half the distance between the nuclei of adjacent atoms in the crystal of a noble gas measures what is known as the van der Waals' radius of the atom concerned. It gives the best numerical value for the actual size of an atom, for it measures how close two atoms will approach when not attracted towards each other by any very strong bond. The values for the noble gas atoms (in nm) are as follows:

Ne	Ar	Kr	Xe
0·16	0·19	0·20	0·22

Similarly, van der Waals' radii for other atoms can be obtained by measurements on molecular crystals. For a crystal, such as that of iodine, containing diatomic, covalent molecules the bond distance within each molecule will be twice the covalent radius (page 133). But the closest distance between the nuclei of atoms in adjacent molecules will be equal to twice the van der Waals' radius. Because this represents the distance between two almost unbound atoms, van der Waals' radii are invariably greater than the covalent radii which gives the distance between two atoms held together by a bond. van der Waals' radii have values, in fact, very close to those of the corresponding ionic radii (page 53). Some values, in nm, are collected together below:

	a	b		a	b	c		a	b	c
							H	0·10	0·030	0·154
N	0·15	0·074	O	0·140	0·074	0·140	F	0·135	0·072	0·136
P	0·19	0·110	S	0·185	0·104	0·184	Cl	0·180	0·099	0·181
As	0·20	0·121	Se	0·200	0·117	0·198	Br	0·195	0·114	0·195
Sb	0·22	0·141	Te	0·220	0·137	0·221	I	0·215	0·133	0·216

(a = van der Waals' radius b = covalent radius c = ionic radius)

Many sets of atom models used for making up molecular models have spherical atoms with radii proportional to the van der Waals' radii of the atoms concerned. The atoms then have segments removed in such a way that both correct bond lengths and bond angles are given when the model atoms are connected together. The scale drawings given in Fig. 101 illustrate the usage.

The dimensions of well-known groups can also be included. A methyl radical, for example, has an effective radius of 0·20 nm and an aromatic ring has a half-thickness of 0·18 nm.

5 Clathrates There are empty spaces inside many crystals between the units of which the crystal is composed, and it is possible for other atoms or molecules, of the right size, to be trapped within these spaces. The resulting products are known as inclusion compounds or as clathrates, from the Latin word, *clathratus*, which means enclosed behind bars. The entrapped particles are referred to as the guests; the major crystalline structure as the host.

Benzene-1,4-diol, $C_6H_4(OH)_2$, has an 'open' crystal structure held together by hydrogen bonds. If it is crystallised from an aqueous solution, in the presence of argon at 40 atmospheres, the resulting crystals contain trapped argon atoms. There is no real chemical bonding between the guest and the host, and the argon can be released by melting or dissolving the clathrate, which has a formula [C₆H₄(OH)₂]₃A. Similar clathrates can be formed with xenon, krypton, hydrogen sulphide, hydrogen chloride, hydrogen cyanide, sulphur dioxide and carbon dioxide as guests within the benzene-1,4-diol host.

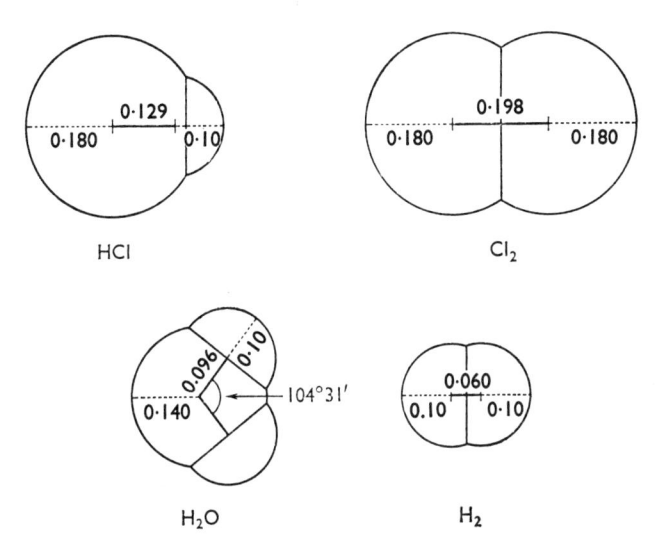

Fig. 101. Simple molecules showing the bond distance (full lines) *and the van der Waals' radii* (dotted lines). *The drawings are to a scale of 1nm = 1cm.*

Benzene molecules can also be trapped by shaking with an ammoniacal solution of nickel(II) cyanide. The resulting clathrate, a pale mauve solid, has a formula Ni(CN)₂.NH₃.C₆H₆. Thiophene will not form such a clathrate, so that clathrate formation provides a method of freeing benzene from thiophene.

Hydrates of the noble gases, e.g. Xe·6H₂O, of methane, e.g. CH₄.5¾H₂O, and of chlorine, e.g. Cl₂.8H₂O are also clathrates, with ice as the host.

15 Complex or Co-ordination Compounds

1 Introduction Many compounds are known which seem to be made up of two perfectly stable molecules combined together for no apparent reason. Potassium hexacyanoferrate(II), for example, has a composition which can be represented by the formula $K_4Fe(CN)_6$, and in the early days of chemistry the only structure which could be allotted to it was written as $Fe(CN)_2 \cdot 4KCN$. That this structure was not satisfactory is shown by the fact that potassium hexacyanoferrate(II) does not form either iron(II) or cyanide ions in solution. Potassium ions are produced, however, and Werner (1866–1919), who first studied such compounds, suggested that the iron and the cyanide radicals were combined together to form what we now call a complex ion. The compound was then written as $K_4[Fe(CN)_6]$, the square bracket indicating that the group within it ionised as a whole, i.e. was a complex ion.

Many similar complex ions are now known. The following are typical examples:

Hexacyanoferrate(II) ion, $Fe(CN)_6^{4-}$
Hexacyanoferrate(III), $Fe(CN)_6^{3-}$
Tetramminecopper(II), $Cu(NH_3)_4^{2+}$
Hexafluorosilicate(IV) ion, SiF_6^{2-}
Hexachlorostannate(IV) ion, $SnCl_6^{2-}$
Pentacyanonitrosylferrate(III) ion, $Fe(CN)_5NO^{2-}$
Hexanitritocobaltate(III) ion, $Co(NO_2)_6^{3-}$
Hexafluoroaluminate ion, AlF_6^{3-}

Salts containing these, and other similar complex ions, are known as *complex salts* or *co-ordination compounds*.

A complex ion is itself made up of a central ion surrounded by other ions or molecules, which are known as *ligands*, the number of ligands round the central ion being called the *co-ordination number*. It is usually 2 or 4 or 6, but may be 8 or an odd number in rare cases. Care must be taken to avoid confusing the use of this term co-ordination number with the crystallographic use of the same term (page 55).

The majority of co-ordination compounds involve complex ions, and, as will be seen, these are formed particularly by transition metals. Uncharged complexes do, however, also exist. Carbonyls, e.g. $Ni(CO)_4$ and $Cr(CO)_6$, are typical examples (page 206) and a series of hydrocarbon complexes, e.g. $Cr(C_6H_6)_2$, $Co_2(CO)_6(C_2H_2)$ and $Fe(C_5H_5)_2$, are of particular modern development and interest (page 207). Neutral complexes can also arise when different groups within a complex neutralise each other's charge, e.g. $Co(NH_3)_3Cl_3$ in which the Co^{3+} charge is neutralised by the Cl^-_3 charge.

2 Chelation Simple or *monodentate* ligands are attached to the central ion by only one atom in the ion or molecule of the ligand. Monodentate ligands are generally anions or neutral molecules, the commonest examples being listed below:

CN^-, Cyano.	OH^-, Hydroxo.	$O.NO^-$, Nitrito.	NO_2^-, Nitro.
F^-, Fluoro.	Cl^-, Chloro.	Br^-, Bromo.	I^-, Iodo.
H_2O, Aquo.	NH_3, Ammine.	CO, Carbonyl.	NO, Nitrosyl.

Other ligands make use of two or more atoms to form more than one bond to the central ion. Such ligands are called chelate groups, and the compounds formed, chelate compounds (from the Greek χηλή = a crab's claw). A group capable of forming two bonds is called a bidentate group; one forming three bonds is a tridentate group; and so on.

(a) Bidentate groups. Ethylenediamine (ethane-1,2-diamine) usually abbreviated as en, is a typical bidentate group, giving typical chelate compounds as shown:

```
      H   H
      |   |
   H—C—N—H
      |   ..
      |   ..
   H—C—N—H
      |   |
      H   H
```
Ethylenediamine
(Ethane-1,2-diamine)

$\{en \supset Cr \subset en / Cl \quad Cl\}^+ Cl^-$ $\{en \supset Co \subset en / () / en\}^{3+} Cl_3^-$

Typical chelate compounds involving ethylenediamine

Other common bidentate groups are acetylacetone, 2,2′-bipyridine, dimethylglyoxime, and 1,10-phenanthroline

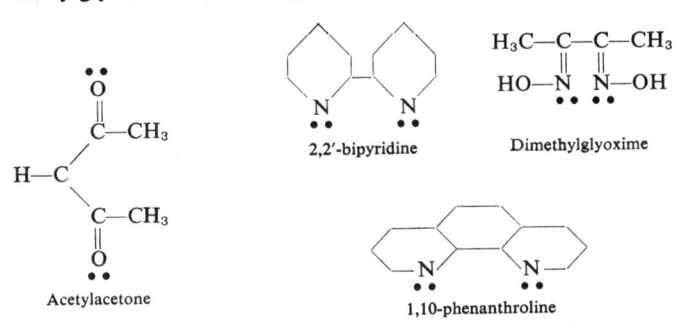

Acetylacetone

2,2′-bipyridine

Dimethylglyoxime: $H_3C-C-C-CH_3$ / $HO-N\ \ N-OH$

1,10-phenanthroline

(b) Tridentate groups. Triamino propane and tripyridine are typical tridentate groups:

```
CH_2.NH_2
|
CH.NH_2
|
CH_2.NH_2
```
Triaminopropane

Tripyridine

(c) *Hexadentate group.* Ethylenediaminetetracetic acid, EDTA, is an important hexadentate group for it forms particularly stable complexes with, for example, Ca^{2+} ions.

$$\begin{array}{l} CH_2\text{—}\ddot{N}\text{=}(CH_2.C\ddot{O}OH)_2 \\ | \\ CH_2\text{—}\ddot{N}\text{=}(CH_2.C\ddot{O}OH)_2 \end{array}$$ Ethylenediaminetetracetic acid (EDTA).

It will be seen that the various ligands contain one or more atoms with lone-pairs of electrons, and these atoms play an important part in the bonding in complexes. Lone-pair electrons are not, however, essential. Hydrocarbons, for example, do not contain atoms with lone-pairs.

3 Nomenclature of complexes The systematic nomenclature for complex ions is based on the Stock-Werner system and is recommended by the International Union of Pure and Applied Chemistry, though it is certainly not always used. The system, in its simplest aspects, is based on the following considerations:

(a) *Complex cations.* The name to be used begins by giving the number and names of the groups attached to the central atom or ion, i.e. of the ligands, and this is followed by the name of the central atom with its oxidation number (page 127) indicated by Roman numerals in parentheses.

The $Cu(NH_3)_4^{2+}$ ion, for example, is called the tetramminecopper-(II) ion, and $Co(NH_3)_4Cl_2^+$ is the dichlorotetramminecobalt(III) ion. Other examples are,

$Ag(NH_3)_2^+$ Diamminesilver (I) $CrCl_2(H_2O)_4^+$ Dichlorotetraquochromium (III)
$Cu(H_2O)_4^{2+}$ Tetraquocopper (II) $Ni(NH_3)_6^{2+}$ Hexamminenickel (II)

(b) *Complex anions.* The name used gives the number and names of the ligands, followed by the name of the central atom with an -ate ending (or an -ic ending for an acid) and its oxidation number in parentheses. Typical examples are:

$CrCl_4(H_2O)_2^-$ Tetrachlorodiaquochromate (III)
$Au(CN)_2^-$ Dicyanoaurate (I)
$Fe(CN)_6^{4-}$ Hexacyanoferrate (II)
$Fe(CN)_6^{3-}$ Hexacyanoferrate (III)
$PtCl_6^{2-}$ Hexachlorplatinate (IV)
$Zn(OH)_4^{2-}$ Tetrahydroxozincate (II)

4 Bonding in complexes The earliest ideas of bonding in complexes were developed by Sidgwick. Dative bonds were envisaged as joining the ligands to the central ion, the lone-pairs from the ligand atoms donating the necessary electrons. The hexacyanoferrate(II) and tetramminecopper(II) ions were represented, for example:

$$\left\{ \begin{array}{c} NC \searrow \underset{\downarrow}{CN} \swarrow CN \\ Fe \\ NC \nearrow \underset{CN}{\uparrow} \nwarrow CN \end{array} \right\}^{4-} \qquad \left\{ \begin{array}{c} NH_3 \\ \downarrow \\ H_3N \rightarrow Cu \leftarrow NH_3 \\ \uparrow \\ NH_3 \end{array} \right\}^{2+}$$

The use of dative bonding in this way could account for the formation of many complexes and the bond was also known by its alternative name of *co-ordinate bond*. The directional nature of the dative bond accounted for the existence of isomers of some complexes. But the earlier work done by Werner (page 177) was soon expanded by the use of X-ray and other physical methods of finding the shapes of molecules.

As experimental information grew, Sidgwick's theories did not provide a satisfactory explanation of the variety of geometrical shapes which emerged, and Pauling's ideas of hybridisation were widely and successfully developed. Later, as will be seen, these ideas were replaced, at least partially, by the ideas of crystal field and ligand field theory.

HYBRIDISATION IN COMPLEXES

5 Introduction Pauling, in the years following 1931, was able to account for, and even predict, the geometrical shape of many complexes on the assumption that hybrid bonds were involved in their structures. The formation of hybrid bonds from hybrid orbitals has been described on pages 107–14, and Pauling related bond hybridisation and geometrical shape according to the following summary (see, also, page 114):

Bond hybridisation	d^2sp^3	sp^3d^2	sp^3	dsp^2
Geometrical shape	Octahedral	Octahedral	Tetrahedral	Square
Typical example	$Fe(CN)_6^{3-}$	FeF_6^{3-}	$Zn(NH_3)_4^{2+}$	$Ni(CN)_4^{2-}$

As will be seen, Pauling made extensive use of magnetic measurements to discover the number of unpaired electrons in a complex (page 64), and he also differentiated between complex ions in which the bonding could be considered as ionic and those in which it was probably covalent.

6 Octahedral complexes. sp^3d^2 **or** d^2sp^3 **hybridisation** Octahedral complexes result from either sp^3d^2 or d^2sp^3 hybrid bonding as in the following examples:

(a) *The hexacyanoferrate(III) ion,* $Fe(CN)_6^{3-}$. The structures of the Fe atom and the Fe^{3+} ion are given at (i) and (ii) below:

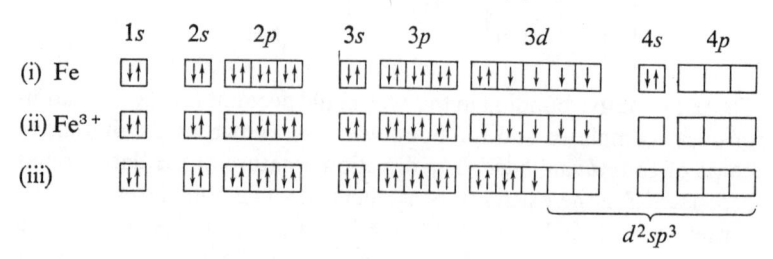

	1s	2s	2p	3s	3p	3d	4s	4p
(i) Fe	↑↓	↑↓	↑↓ ↑↓ ↑↓	↑↓	↑↓ ↑↓ ↑↓	↑↓ ↑ ↑ ↑ ↑	↑↓	☐ ☐ ☐
(ii) Fe^{3+}	↑↓	↑↓	↑↓ ↑↓ ↑↓	↑↓	↑↓ ↑↓ ↑↓	↑ ↑ ↑ ↑ ↑	☐	☐ ☐ ☐
(iii)	↑↓	↑↓	↑↓ ↑↓ ↑↓	↑↓	↑↓ ↑↓ ↑↓	↑↓ ↑↓ ↑ ☐ ☐	☐	☐ ☐ ☐

$\underbrace{\qquad\qquad\qquad\qquad}_{d^2sp^3}$

The five unpaired electrons in the Fe^{3+} ion give a calculated magnetic moment (page 64) for iron(III) compounds of 5·92 magneton, which compares with a measured value for iron(III) sulphate of 5·86.

The magnetic moment for the $Fe(CN)_6^{3-}$ ion, however, is 2·3 suggesting, perhaps, one unpaired electron, though the calculated magnetic moment for one electron is 1·73.

It is suggested that the unpaired 3d-electrons in the Fe^{3+} ion couple as fully as possible, prior to bond formation, to give the arrangement shown at (iii). The resulting complex would have one unpaired electron, and the d^2sp^3-orbitals available for hybridisation are bracketed together. The electrons to occupy the d^2sp^3 hybrid orbitals would be provided by the lone pairs from each of the six cyanide ion ligands. This means, essentially, that the bonds between the ligands and the central ion are being portrayed as dative bonds, but such an idea must be an oversimplification for six dative bonds would lead to the development of such a high charge on the central ion (see page 204).

Complexes like $Fe(CN)_6^{3-}$ are known as *inner-orbital complexes* because the *d*-orbitals involved are from a lower shell than the *s*- and *p*-orbitals.

(b) *The hexacyanoferrate(II) ion,* $Fe(CN)_6^{4-}$. In the hexacyanoferrate(III) ion there is one more electron than in the hexacyanoferrate (III) ion. This extra electron pairs with the unpaired 3d-electron in (iii), above, to give the structure shown at (iv) below. The hexacyano-

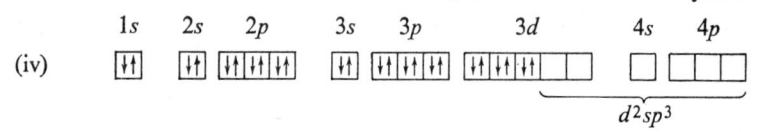

	1s	2s	2p	3s	3p	3d	4s	4p
(iv)	↑↓	↑↓	↑↓ ↑↓ ↑↓	↑↓	↑↓ ↑↓ ↑↓	↑↓ ↑↓ ↑↓ ☐ ☐	☐	☐ ☐ ☐

$\underbrace{\qquad\qquad\qquad\qquad}_{d^2sp^3}$

ferrate(II) ion has, therefore, no unpaired electrons and would be expected to be diamagnetic; this is found to be so. It is octahedral, with d^2sp^3 hybridisation.

Other complexes with structures like the hexacyano-ferrate(II) or ferrate(III) ions include $Co(NH_3)_6^{3+}$, $Mn(CN)_6^{3-}$, $Cr(NH_3)_6^{3+}$ and $Cr(CN)_6^{4-}$.

(c) *The hexafluoroferrate(III) ion*, FeF_6^{3-}. The hexafluoroferrate(III) ion is octahedral, like the hexacyano-ferrate(II) or ferrate(III) ions, but it has a magnetic moment of 6·0, in close agreement with the value of 5·92 which would be expected from a structure with five unpaired electrons.

To account for this, it is suggested that no pairing of electron spins takes place in the Fe^{3+} ion and that sp^3d^2 orbitals, as shown bracketed together below:

1s 2s 2p 3s 3p 3d 4s 4p 4d

[↑↓] [↑↓][↑↓↑↓↑↓] [↑↓][↑↓↑↓↑↓] [↑↑↑↑↑] [ ][ ][ ][ ]
                                              $\underbrace{}_{sp^3d^2}$

are used for hybridisation. Because the free Fe^{3+} ion also has five unpaired electrons, it has been suggested that the bonding in the FeF_6^{3-} ion is ionic.

The FeF_6^{3-} ion is an example of an *outer-orbital complex* because the d-orbitals involved in the hybridisation are from a higher shell than the s- and p-orbitals.

Other similar outer-orbital complexes include CoF_6^{3-}, $Fe(NH_3)_6^{2+}$, $Ni(NH_3)_6^{2+}$, $Cu(NH_3)_6^{+}$ and $Cr(H_2O)_6^{2+}$.

7 Tetrahedral complexes. sp^3 **hybridisation** Complexes of Zn^{2+} are invariably tetrahedral because they involve sp^3-hybrid bonds. The arrangements of electrons in the Zn atom and the Zn^{2+} ion are shown at (i) and (ii) below:

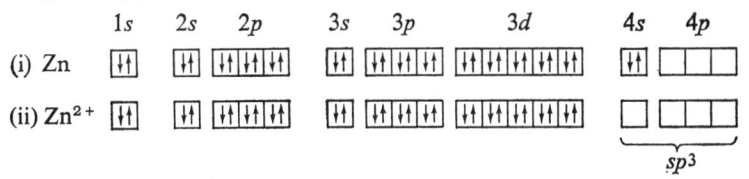

       1s 2s 2p 3s 3p 3d 4s 4p
(i) Zn [↑↓] [↑↓][↑↓↑↓↑↓] [↑↓][↑↓↑↓↑↓] [↑↓↑↓↑↓↑↓↑↓] [↑↓][ ][ ][ ]

(ii) Zn^{2+} [↑↓] [↑↓][↑↓↑↓↑↓] [↑↓][↑↓↑↓↑↓] [↑↓↑↓↑↓↑↓↑↓] [ ][ ][ ][ ]
                                                                $\underbrace{}_{sp^3}$

Because all the 3d-orbitals are fully occupied they cannot participate in bond hybridisation. Four sp^3-hybrid orbitals can be formed, however, as indicated by the bracket and the resulting complexes, e.g. $Zn(NH_3)_4^{2+}$, will be tetrahedral. As there are no unpaired electrons they will also be diamagnetic.

Other similar tetrahedral complexes are formed by Cd^{2+}, Hg^{2+}, Cu^+ and Ag^+ ions, all of which have the same structure as Zn^{2+}.

8 Square planar complexes. *dsp^2 hybridisation* Hybridisation of a *d*-, an *s*- and two *p*-orbitals gives four equivalent hybrid orbitals which are planar and mutually at right-angles pointing towards the corners of a square. The *d*-orbital involved is the $d_{x^2-y^2}$-orbital in the shell lower than that of the *s*- and *p*-orbitals.

Square complexes tend to be formed when the central ion has only one *d*-orbital available in the lower shell.

(*a*) *The tetracyanonickelate(II) ion*, $Ni(CN)_4^{2-}$. The electronic structures for the Ni atom and the Ni^{2+} ion are shown at (i) and (ii) below:

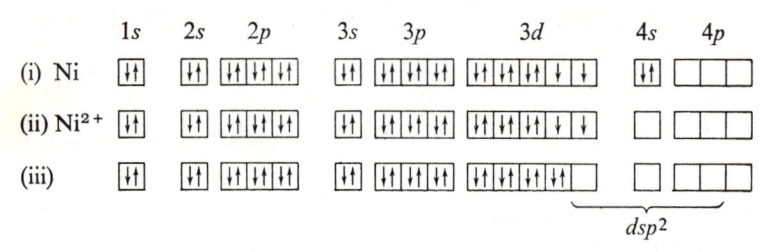

	1s	2s	2p	3s	3p	3d	4s	4p
(i) Ni	↑↓	↑↓	↑↓ ↑↓ ↑↓	↑↓	↑↓ ↑↓ ↑↓	↑↓ ↑↓ ↑↓ ↑ ↑	↑↓	□ □ □
(ii) Ni^{2+}	↑↓	↑↓	↑↓ ↑↓ ↑↓	↑↓	↑↓ ↑↓ ↑↓	↑↓ ↑↓ ↑↓ ↑ ↑	□	□ □ □
(iii)	↑↓	↑↓	↑↓ ↑↓ ↑↓	↑↓	↑↓ ↑↓ ↑↓	↑↓ ↑↓ ↑↓ ↑↓ □	□	□ □ □

$$\underbrace{\qquad\qquad\qquad\qquad}_{dsp^2}$$

If the two unpaired 3*d*-electrons couple, the structure will be as at (iii) so that dsp^2-hybrid bonds can be formed. All the electrons are paired so that the $Ni(CN)_4^{2-}$ ion is diamagnetic. Other similar examples are provided by $PtCl_4^{2-}$, $PdCl_4^{2-}$.

(*b*) *The tetramminecopper(II) ion*, $Cu(NH_3)_4^{2+}$. This common complex ion, which occurs in Schweitzer's solution, has a co-planar, square structure and is paramagnetic with a magnetic moment of 1·8. A square structure necessitates dsp^2 hybridisation, and the magnetic moment suggests one unpaired electron.

Such facts can be accounted for on the basis of the structures shown at (i), (ii) and (iii) below:

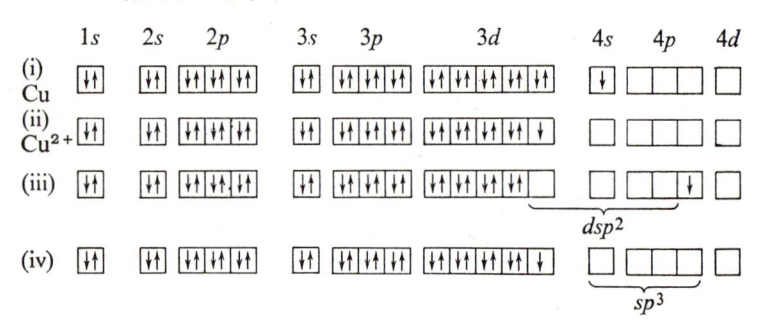

	1s	2s	2p	3s	3p	3d	4s	4p	4d
(i) Cu	↑↓	↑↓	↑↓ ↑↓ ↑↓	↑↓	↑↓ ↑↓ ↑↓	↑↓ ↑↓ ↑↓ ↑↓ ↑↓	↓	□ □ □	□
(ii) Cu^{2+}	↑↓	↑↓	↑↓ ↑↓ ↑↓	↑↓	↑↓ ↑↓ ↑↓	↑↓ ↑↓ ↑↓ ↑↓ ↑	□	□ □ □	□
(iii)	↑↓	↑↓	↑↓ ↑↓ ↑↓	↑↓	↑↓ ↑↓ ↑↓	↑↓ ↑↓ ↑↓ ↑↓ □	□	□ □ ↓	□

$$\underbrace{\qquad\qquad\qquad}_{dsp^2}$$

| (iv) | ↑↓ | ↑↓ | ↑↓ ↑↓ ↑↓ | ↑↓ | ↑↓ ↑↓ ↑↓ | ↑↓ ↑↓ ↑↓ ↑↓ ↑ | □ | □ □ □ | □ |

$$\underbrace{\qquad\qquad\qquad}_{sp^3}$$

The necessity to promote one electron from a 3*d* into a 4*p* level is not, however, entirely satisfactory. The 4*p*-electrons would be expected to be easily lost which would mean that the complex could be easily oxidised, but this is not so.

Placement of the odd electron in the 3*d* level, as in (iv), would suggest sp^3 hybridisation with a resulting non-existent tetrahedral structure.

The matter is probably best resolved by regarding this complex as having the structure of a distorted octahedron, as explained on page 197.

STABILITY OF COMPLEXES

The term 'stability' can be used in a number of different ways, and it is commonly used rather vaguely. So far as complexes are concerned it is not easy to use the term in any generalisations for a complex may be quite stable to one reagent and yet decompose readily in reaction with another. The term may also refer to the action of heat or light on a compound.

Nevertheless, a number of important topics, many of them contemporary research projects, can be collected together under the general umbrella of stabilisation.

9 Heat of hydration The heat of formation of a complex, e.g.

$$M^{2+} + 6L \to ML_6^{2+} + \text{Heat of formation}$$

gives a thermodynamic measure of the stability of the complex, and interesting results have been obtained from measurements of the heats of hydration of the divalent ions between Ca^{2+} and Zn^{2+}, e.g.

$$Cr^{2+} + 6H_2O \to Cr(H_2O)_6^{2+} + \text{Heat of hydration}$$

The results obtained are summarised, graphically, in Fig. 102. There are maxima for V^{2+} and Ni^{2+} and a minimum for Mn^{2+}, suggesting that $V(H_2O)_6^{2+}$ and $Ni(H_2O)_6^{2+}$ are the most stable of these hydrates and that $Mn(H_2O)_6^{2+}$ is the least stable.

Such results are unexpected. The radii of the divalent ions falls steadily in passing from Ca^{2+} to Zn^{2+} and the electrostatic fields around the ions would, on these grounds, be expected to increase. A steady rise in the stability of $M(H_2O)_6^{2+}$ complexes in passing from $Ca(H_2O)_6^{2+}$ to $Zn(H_2O)_6^{2+}$ would, therefore, be expected. So far as the hydrates of the alkaline-earths are concerned, there is a steady rise in passing from Ba^{2+} to Mg^{2+}, i.e. the heat of hydration does increase as the ionic radius gets smaller. But the heats of hydration of the trivalent ions between Sc^{3+} and Ga^{3+} also show discontinuities, with a pronounced minimum at Fe^{3+} (Fig. 103).

Such anomalous results for measured heats of hydration of transi-

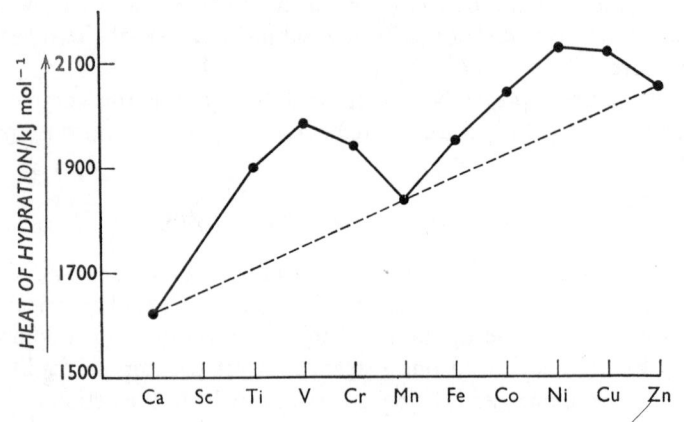

Fig. 102. *The heats of hydration of various divalent ions. The dotted line refers to the corrected values (see page 191)*

tional metal ions cannot be accounted for in terms of hybridisation, but the ideas of crystal field theory provide a neat solution to the problem, as explained on page 191.

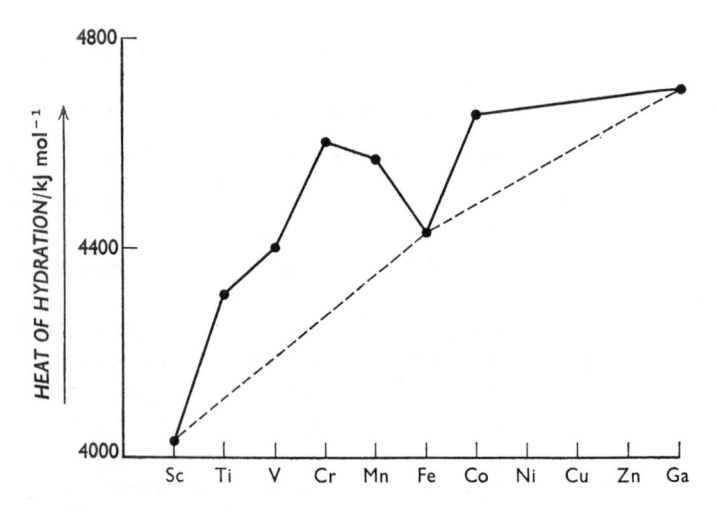

Fig. 103. *The heats of hydration of various trivalent ions. The dotted line refers to the corrected values (p. 191)*

10 Stability of complexes in aqueous solution When ammonia is added to an aqueous solution of, say, nickel(II), sulphate, $NiSO_4$, there is a noticeable colour change because some of the $Ni(H_2O)_6^{2+}$ ions are converted into $Ni(NH_3)_6^{2+}$ ions. The reaction taking place is a replacement reaction involving ligands.

$$Ni(H_2O)_6^{2+} + 6NH_3 \rightleftharpoons Ni(NH_3)_6^{2+} + 6H_2O \qquad (I)$$

The resulting mixture would contain both $Ni(H_2O)_6^{2+}$ and $Ni(NH_3)_6^{2+}$ ions and the equilibrium constant for the reaction,

$$\frac{[Ni(NH_3)_6^{2+}] \cdot [H_2O]^6}{[Ni(H_2O)_6^{2+}] \cdot [NH_3]^6} = \text{Equilibrium constant}$$

would give a measure of the relative stabilities of the two ions. A high value would mean that the $Ni(NH_3)_6^{2+}$ ion was very stable, i.e. that the equilibrium lay well over to the right.

In aqueous solution, the concentration of water can be regarded as constant, for it changes so little, so that

$$\frac{[Ni(NH_3)_6^{2+}]}{[Ni(H_2O)_6^{2+}] \cdot [NH_3]^6} = K$$

would have a constant value, K. This is known as the *stability constant* for the $Ni(NH_3)_6^{2+}$ ion; its numerical value is $9 \cdot 98 \times 10^7$. The higher the stability constant for a complex ion the greater its stability. Alternatively, $1/K$ values, sometimes called *instability constants*, may be quoted. They give a measure of the extent to which the equilibrium represented in the following equation, which is the reverse of (I) above, lies to the right,

$$Ni(NH_3)_6^{2+} + 6H_2O \rightleftharpoons Ni(H_2O)_6^{2+} + 6NH_3$$

A high value for an instability constant means an unstable complex ion.

Stability constants are often quoted as lg K, and some typical values are given below:

$Co(NH_3)_6^{2+}$ 5·11 $Ni(NH_3)_6^{2+}$ 7·99
$Zn(NH_3)_4^{2+}$ 9·1 $Cu(NH_3)_4^{2+}$ 12·6
$Fe(CN)_6^{4-}$ 8·3 $Fe(CN)_6^{3-}$ 31·0

The stability constants for ions such as $Ni(NH_3)_6^{2+}$ are made up of a number of terms, for the reaction indicated by the equation at (I) takes place in six stages, one ligand being replaced at a time. Each of the six stages has its own equilibrium constant and a corresponding value for the stability constant for the ion concerned:

$Ni(H_2O)_6^{2+}$ $+ NH_3 \rightleftharpoons Ni(H_2O)_5NH_3^{2+}$ $+ H_2O \ldots K_1$
$Ni(H_2O)_5NH_3^{2+}$ $+ NH_3 \rightleftharpoons Ni(H_2O)_4(NH_3)_2^{2+} + H_2O \ldots K_2$
$Ni(H_2O)_4(NH_3)_2^{2+} + NH_3 \rightleftharpoons Ni(H_2O)_3(NH_3)_3^{2+} + H_2O \ldots K_3$
$Ni(H_2O)_3(NH_3)_3^{2+} + NH_3 \rightleftharpoons Ni(H_2O)_2(NH_3)_4^{2+} + H_2O \ldots K_4$
$Ni(H_2O)_2(NH_3)_4^{2+} + NH_3 \rightleftharpoons Ni(H_2O)(NH_3)_5^{2+} + H_2O \ldots K_5$
$Ni(H_2O)(NH_3)_5^{2+}$ $+ NH_3 \rightleftharpoons Ni(NH_3)_6^{2+}$ $+ H_2O \ldots K_6$

The overall stability constant, K, is equal to the product of the successive stability constants, so that

$$K = K_1 \times K_2 \times K_3 \times K_4 \times K_5 \times K_6$$

Alternatively,

$$\lg K = \lg K_1 + \lg K_2 + \lg K_3 + \lg K_4 + \lg K_5 + \lg K_6$$

The following figures for $Ni(NH_3)_6^{2+}$ illustrate the point:

K_1	K_2	K_3	K_4	K_5	K_6	K
5×10^2	$1 \cdot 3 \times 10^2$	40	12	4	0·8	$9 \cdot 98 \times 10^7$
$\lg K_1$	$\lg K_2$	$\lg K_3$	$\lg K_4$	$\lg K_5$	$\lg K_6$	$\lg K$
2·70	2·11	1·60	1·08	0·60	−0·10	7·99

It is a general rule that the successive stability constants decrease regularly in passing from K_1 to K_6. A log K value greater than 8 is sometimes taken as meaning a stable complex, though the matter is relative and the choice of figure arbitrary.

11 Effect of central ion on stability of complexes Efforts to systematise the mass of data regarding the stability of complexes have been only partially successful. So many factors seem to be involved that it is difficult to isolate them. Some of the factors seem to be mainly dependent on the central ion; others on the ligand (page 173). In reality it is no doubt the interplay of the two sets of factors which controls the situation.

(a) *Ionic size.* The smaller an ion, the greater its electrical field, and, in general, the more stable its complexes. The following stability constants for hydroxide complexes of the alkaline-earths illustrate the point:

	$BeOH^+$	$MgOH^+$	$CaOH^+$	$BaOH^+$
$K =$	10^7	120	30	4

(b) *Ionic charge.* For ions of equal size, those with higher charge exert the stronger field and form the stablest complexes. The hexacyanoferrate(III) ion ($\lg K = 31 \cdot 0$), for example, is more stable than the hexacyanoferrate(II) ($\lg K = 8 \cdot 3$), and cobalt(II) complexes are more stable than those of cobalt(III) (page 175).

Combination of (a) and (b) shows that it is the charge/radius ratio which is significant. A large charge/radius ratio leads to stable complexes.

(c) *Electronegativity.* The bonding between a ligand and a central ion is, to some extent, due to the donation of electrons by the ligand to the central ion. It might be expected, then, that a strongly electron-attracting central ion, i.e. one with a high electronegativity, would form the stablest complexes. This is, in some ways, true, but the relationship is by no means direct.

For metals which have high electropositivity, i.e. the alkali metals and the alkaline-earths, sometimes called *class a metals*, the stablest complexes are formed with strong electron donors (page 204). But for metals with higher electronegativities, sometimes classified as *b metals*, the electron-attracting powers of the central atom seem to be of lesser significance.

(*d*) *The natural order of stability.* Despite all the diffculties it is possible to list a limited number of ions in order of the stability of the complexes they form:

	Mn^{2+}	< Fe^{2+}	< Co^{2+}	< Ni^{2+}	< Cu^{2+}	> Zn^{2+}
Ionic radius/nm	0·091	0·083	0·082	0·078	0·069	0·074

For this limited number of ions the order is almost independent of the ligand used in complex formation. The stability constants (lg K) for the ethylenediamine and EDTA complexes illustrate the point:

	Mn^{2+}	Fe^{2+}	Co^{2+}	Ni^{2+}	Cu^{2+}	Zn^{2+}
$M(en)_3^{2+}$	5·67	9·52	13·82	18·06	18·60	12·09
$M(EDTA)^{2+}$	13·8	14·2	16·1	18·5	18·8	16·5

It will be seen that the order of these six ions, sometimes known as the natural order of stability, is the order of their ionic radii.

The numerous attempts to extend the very limited list given above, and to draw up similar lists for trivalent ions, have not been very successful. Lists can be drawn up which apply very well for one ligand, but the lists are not comprehensive enough to be of much value.

12 Effect of ligand on stability of complexes Just as it would be satisfying and useful to list simple ions in their correct 'stability' order irrespective of the ligands used, so it would be helpful if ligands could be listed irrespective of the central ion. But, again, this cannot be done at all fully, and many factors are involved.

(*a*) *Size and charge of ligand.* For charged ligands the higher the charge and the smaller the size, the more stable the complex formed. The F^- ion, for example, forms stabler complexes than the larger Cl^- ion. Compare, for instance, the stability constants of FeF^{2+} (1×10^6) and $FeCl^{2+}$ (20).

(*b*) *Basic character.* The basicity of a ligand is an important factor for this depends on how easily it can donate electrons. A strong base is a good electron donor, and many common ligands, e.g. CN⁻, F⁻ and NH₃, are also strong bases. The formation of an acid and a complex have, indeed, something in common,

$$H^+ + Base \rightleftharpoons Acid$$
$$M^+ + Ligand \rightleftharpoons Complex$$

This analogy is particularly close when the complexes of highly electropositive metals, i.e. class *a* metals, are considered. In such cases M^+ and H^+ have much in common.

(c) *The extent of chelation.* Multidentate ligands, unless they are very large (see (d) below) form stabler complexes than monodentate ones. The stability constants ($\log_{10} K$) for the ammonia, ethylene diamine and triethylene tetramine

$$(NH_2CH_2CH_2NHCH_2CH_2NHCH_2CH_2NH_2)$$

complexes of the copper(II) ion illustrate the point

	$Cu(NH_3)_4^{2+}$	$Cu(en)_2^{2+}$	$Cu(trien)^{2+}$
lg K	11·9	20·0	20·5

The figures quoted in section **11** also show that the EDTA complex is always stabler than that with ethylene diamine. It is, indeed, the stability of the complexes which it will form that makes EDTA so useful.

(d) *Steric effects.* Large, bulky ligands form less stable complexes than similar smaller ones. The ethylene diamine complexes, for example, are stabler than those of the corresponding tetramethyl-substituted derivative, $(CH_3)_2NCH_2CH_2N(CH_3)_2$.

(e) *Size of chelate ring.* The stablest complexes are formed by saturated ligands that form five-membered chelate rings or by unsaturated ligands forming six-membered rings.

13 Stabilisation of unstable valency states The formation of a complex often stabilises an otherwise unstable ion, generally by making it more resistant to oxidation or reduction. Some typical examples are given below:

(a) *Cu(en)₂I₂.* Copper(II) iodide is unstable. It splits up into copper(I) iodide and iodine, and use is made of this in the common iodimetric method of estimating copper(II) ions. The complex, Cu(en)$_2$I$_2$, is, however, quite stable with no tendency to be reduced to the copper(I) state.

(b) *Copper(I) complexes.* Copper(I), Cu$^+$, ions change, in aqueous solution, into copper and copper(II), Cu^{2+}, ions,

$$2Cu^+(aq) \rightarrow Cu + Cu^{2+}(aq)$$

In other words, Cu^{2+} ions, under these conditions, are more stable than Cu$^+$ ions; the equilibrium constant for the above reaction is $1\cdot2 \times 10^6$. In the reaction, one Cu$^+$ ion is converted into a Cu atom (with a lower oxidation state) whilst the other Cu$^+$ ion is converted into a Cu^{2+} ion (with a higher oxidation state); a change of this nature is known as *disproportionation.*

Complexes of Cu$^+$ ions, e.g. Cu(CN)$_4^{3-}$ and Cu(NH$_3$)$_4^+$, are, however, stable in aqueous solution, as are insoluble copper(I) compounds, e.g. copper(I) chloride, iodide and cyanide.

(c) *Cobalt(III) complexes.* Simple cobalt(III) salts are difficult to make, and unstable when made, because it is difficult to oxidise Co^{2+} to Co^{3+} even using electrolytic methods. The Co(NH$_3$)$_6^{3+}$ and Co(CN)$_6^{3-}$ ions are, however, very stable. So much so, in fact, that even air will oxidise Co^{2+} to Co^{3+} in the presence of cyanide ions or ammonia, and the resulting cobalt(III) complexes are much more stable than the corresponding cobalt(II) ones.

This may well be due to the fact that cobalt(III) complexes involving d^2sp^3 hybridisation would have the 3d-, 4s- and 4p-orbitals just filled, whereas the cobalt(II) complexes would have to have one electron in the higher 5s level if they were to form d^2sp^3-hybrid bonds. The position is summarised at (i) and (ii) below:

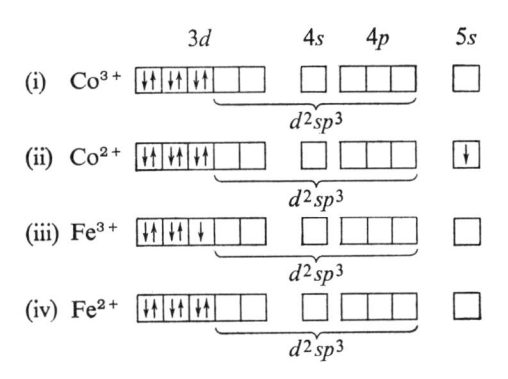

Iron has one less electron than cobalt so that there is no necessity to promote an electron into the 5s level in either iron(III) or iron(II) complexes. Both sets of complexes are, in fact, reasonably stable. The structures are shown above, at (iii) and (iv).

Ni^{3+} has the same structure as Co^{2+}, so that it is not surprising that octahedral NiX_6^{3+} complexes are very rare.

(d) *Complexes of iron.* Many simple iron(II) compounds are fairly readily oxidised, sometimes on exposure to air, but the complexes formed by iron(II) ions with, for example, phenanthroline (page 163) are much more stable. *o*-phenanthroline iron(II) sulphate, ferroin, which is orange-red in solution, can only be oxidised to the iron(III) state, which is pale blue in solution, by strong oxidising agents. It is useful as a redox indicator.

Complexing with phenanthroline stabilises the divalent state of iron, but complex formation with cyanide ions stabilises the trivalent state. The extent of such stabilisations when oxidation-reduction reactions are involved, can be measured in terms of redox potentials, the following values illustrating the point:

$$Fe(H_2O)_6^{2+} \rightleftarrows Fe(H_2O)_6^{3+} + 1e \quad E° = -0.77 \text{ volt}$$

$$Fe(phen)_3^{2+} \rightleftarrows Fe(phen)_3^{3+} + 1e \quad E° = -1.14 \text{ volt}$$

$$Fe(CN)_6^{4-} \rightleftarrows Fe(CN)_6^{3-} + 1e \quad E° = -0.36 \text{ volt}$$

14 Labile and inert complexes The stability constants described in the preceding section give a measure of the final position attained in an equilibrium mixture, but they do not give any indication of the time taken to reach the equilibrium point.

When reaction rates are considered, complexes can be classified into two general types. Labile complexes react rapidly, often in the course of mixing; inert complexes react much more slowly. There is a remarkably clear-cut line of distinction and Taube first suggested that there was a relation between the type of complex and its structure.

Inner-orbital complexes, for example, are invariably labile if they contain less than three *d*-electrons, whilst they are inert with three or more electrons. The position so far as outer-orbital complexes are concerned is not so straightforward, and a full study of the relationships between reaction rates and structure demands a detailed treatment of reaction mechanism. Such a study has been developed, in recent years, particularly by Basolo and Pearson.

INORGANIC ISOMERISM

15 Introduction It was by studying the various isomers which could be isolated that Werner was able to predict the geometrical arrangement of many complex ions. A four-co-ordinated complex, Ma_2b_2, for instance, will exist in two isomeric forms if the arrangement is planar, but only one if it is tetrahedral. Similarly, a six-co-ordinated complex, Ma_4b_2, will have two isomers if the arrangement is octahedral but three (corresponding to ortho, meta, and para substitution in organic chemistry) if it is planar and hexagonal. Thus an experimental determination of the number of isomers formed by a particular complex ion fixes the geometric arrangement within the ion.

The main types of inorganic isomerism are summarised below.

16 Geometric isomerism This type of isomerism occurs in six-co-ordinated complexes with an octahedral arrangement, and in four-co-ordinated complexes with a planar arrangement. It does not occur in tetrahedral complexes.

Thus $[Co(NH_3)_4Cl_2]Cl$ exists in two forms represented as follows:

The cis-form
Blue-violet

The trans-form
Green

and many other Ma_4b_2 and Ma_3b_3 complex ions form similar isomers.

In planar (four-co-ordinated) complexes the two forms of Ma_2b_2 are represented as follows:

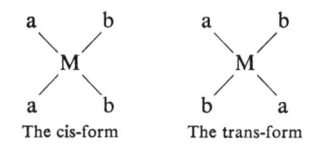

The cis-form   The trans-form

e.g. $Pt(NH_3)_2Cl_2$.

17 Optical isomerism Optical isomerism arises, as in organic chemistry, when a compound can be represented by two asymmetrical structures one being the mirror image of the other. This occurs in many chelate compounds; for example, [Co(en)$_2$Cl$_2$]Cl has optically active isomers represented by

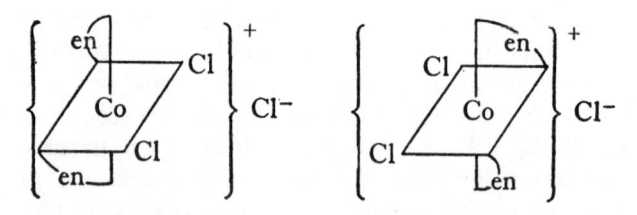

both of which are cis-forms, and also an inactive trans-form:

A similar example which does not contain any carbon atoms is provided by [Rh(SO$_2$N$_2$H$_2$)$_2$(H$_2$O)$_2$]$^-$Na$^+$ the cis-form of which is resolvable into optical isomers. The structures are:

Trans-form

Optical isomers of cis-form

18 Ionisation isomerism This occurs when an ion can occupy a position either inside or outside the complex ion, e.g.

$[Co(NH_3)_5SO_4]Br$ $[Co(NH_3)_5Br]SO_4$
Sulphatopentamminecobalt (III)   Bromopentamminecobalt (III)
bromide   sulphate

The first compound will give a precipitate with silver nitrate(V) solution but not with barium chloride solution; the second compound will behave in the opposite way.

16 Crystal Field Theory

1 Introduction   The use of bond hybridisation, by Pauling, enabled great progress to be made in interpreting the geometrical shape of complexes, but hybridisation, by itself, failed to account for the mass of data which rapidly accumulated.

There were unexplained differences between measured and calculated magnetic moments; compounds were found which had intermediate values for their magnetic moments; the distinction between ionic and covalent bonding in complexes seemed to be too clear-cut; and difficulties arose over the interpretation of geometrical shapes, e.g. the square shape of $Cu(NH_3)_4^{2+}$ (page 168), particularly over distorted shapes. Of still greater importance, Pauling's ideas offered little or no explanation of the absorption spectra of complexes or of their relative stabilities (page 169) or reactivities (page 171).

Such problems have led to a radical revision of ideas concerning the nature of bonding in complexes. This has come about, since 1951, by the application of the ideas of crystal or ligand field theory to chemical problems. The theory was first introduced by Bethe and van Vleck and applied mainly to ionic crystals, but Orgel has been largely responsible for developing its general chemical aspects.

On this theory, a complex is regarded as an agglomeration of a central ion surrounded by other ions or molecules with electrical dipoles. The electrical field of the central ion will affect the surrounding ligands, whilst the combined field of the ligands will influence the electrons of the central ion. As will be seen, the effect of the ligands is particularly marked on the d-electrons which play such a large part in complex formation by transition elements. The influence of the ligands depends on their nature, particularly on the strength of their electrical field, and on their geometrical positions in the complex.

In the simplest treatment of the problems concerned the bonding between the central ion and the ligands is regarded as purely ionic so that the forces involved can be limited to electrostatic forces. Such a limited treatment is generally referred to as *crystal field theory*. A fuller treatment of ligand field theory introduces the possibilities of covalent bonding, as described in Chapter 18.

2 Degeneracy of d-orbitals   In a free ion the five d-orbitals (page 78) are degenerate, i.e. energetically alike, and electrons which occupy such orbitals do so according to the rule of maximum multiplicity (page 28).

Crystal field theory is basically concerned with the effect of different arrangements of surrounding ligands on the energy of the d-orbitals, and the subdivision of the orbitals into two groups, t_{2g} and e_g (pages 78–9), is of very great significance. As will be seen, the d-orbitals are not all alike energetically when under the influence of surrounding ligands. In technical terms, the degeneracy is *resolved* or *split*.

The ligands in the complexes to be considered are either negatively charged ions, e.g. F^- and CN^-, or molecules which have a dipole such that the negative end is closest to the central ion, e.g. H_2O and NH_3. Such ligands will exert an electrostatic field which will tend to repel the electrons, particularly the outer d-electrons, of the central

(a)   (b)   (c)

Free ion, no ligand field   *Crystal field-splitting in an octahedral field*

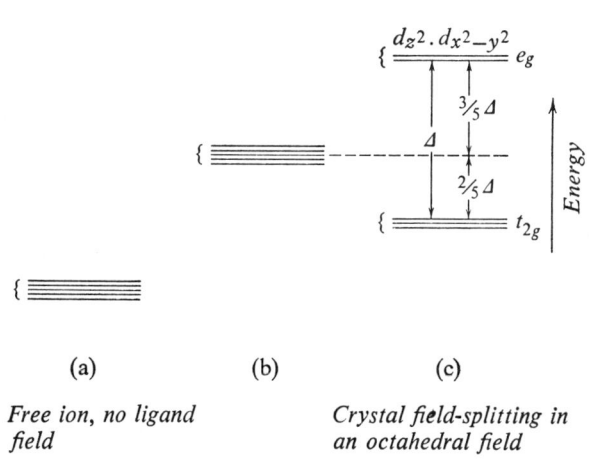

Fig. 104. *The energy levels of d orbitals.* (a) *Degenerate levels in a free ion.* (b) *Degenerate levels at a higher energy level.* (c) *Crystal field splitting in an octahedral field*

ion. This repulsion will raise the energy level of the d-orbitals in the central ion. If the d-orbitals were all alike and the ligand field affected them all in the same way, the five d-orbitals would remain degenerate at a higher energy level (Fig. 104b). In fact, however, the d-orbitals are not all alike; they form two groups. Moreover, the ligand field depends on whether the ligands are arranged in an octahedral, tetrahedral or square planar way around the central ion. As a result the degeneracy of the d-orbitals is resolved or split by the ligand field; the effect is known as *crystal field splitting* (Fig. 104c).

Some ligands exert stronger fields than others, and it is possible to list the common ligands in the order of their field strength in what is known as the *spectrochemical series* (page 189).

3 Crystal field splitting by different geometrical arrangements of ligands The way in which the degeneracy of the *d*-orbitals is resolved is different in octahedral, tetrahedral and square planar fields.

(a) Octahedral field. Figure 105 shows the way in which six ligands are arranged octahedrally around a central ion, and it is clear that the repulsive forces exerted by the ligands will be strongest along the directions of the *x*-, *y*- and *z*-axes.

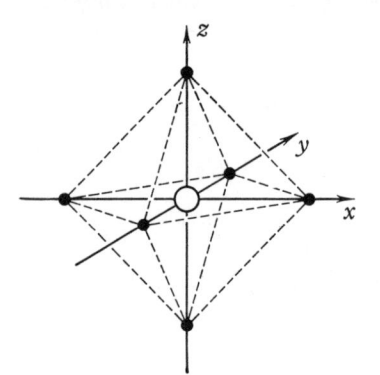

Fig. 105. *The octahedral arrangement of six ligands around a central ion. The ligands lie on the x-, y- and z-axes.*

The $d_{x^2-y^2}$- and the d_{z^2}-orbitals of the central ion are aligned along the *x*- *y*- and the *z*-axes respectively (Fig. 39, page 79), whilst the remaining three *d*-orbitals (the d_{xy}-, d_{xz}- and d_{yz}-orbitals) are directed along lines between the *x*-, *y*- and *z*-axes. It is clear, then, that the effect of the ligands will be greater on the $d_{x^2-y^2}$- and d_{z^2}-orbitals than on the d_{xy}-, d_{xz}- and d_{yz}-orbitals. In other words, *the e_g set of orbitals is affected more than the t_{2g} set.*

As a result, more energy will be required for an electron to occupy an e_g-orbital than a t_{2g}-orbital, and the t_{2g}-orbitals will be occupied before the e_g ones.

The energy difference between the e_g and the t_{2g} orbitals may be represented as shown in Fig. 104c, where the symbol Δ is used to denote the energy difference. The symbol Dq is similarly used. The total energy difference may be regarded as made up of the energy lost in the three lower levels together with the energy gained in the two upper levels. The e_g levels are, therefore, raised by $\frac{3}{5}\Delta$ whilst the t_{2g} levels are lowered by $\frac{2}{5}\Delta$. Actual values of Δ (page 188) vary considerably but values between 120 and 250 kJ mol^{-1} are common.

(b) *Tetrahedral field.* The four ligands arranged tetrahedrally around a central ion are not aligned directly with any of the *d*-orbitals of the ion. The four ligands may be represented as occupying alternate

Fig. 106. *The tetrahedral arrangement of four ligands around a central ion. The directions of the e_g orbitals* (shown as bold lines) *pass through the centres of the cube faces. The directions of the t_{2g} orbitals pass through the centres of the cube edges; only the d_{xy} t_{2g} orbital is shown*

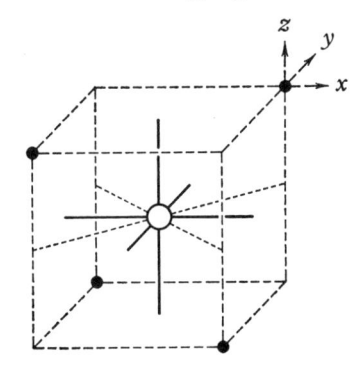

corners of a cube which has the central ion at its centre (Fig. 106). The directions of the e_g-orbitals pass through the centres of the cube faces and are not parallel to the lines joining pairs of ligands. The t_{2g}-orbitals pass through the mid-points of the cube edges and are parallel to the lines joining pairs of ligands across the diagonal of the

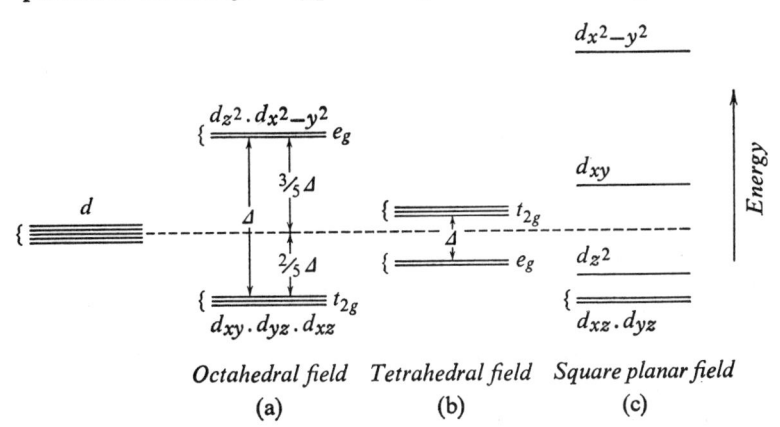

Fig. 107. *Crystal field splitting in different fields*

cube faces. Each lobe of the t_{2g}-orbitals comes closer to a ligand than the lobes of the e_g-orbitals. *In a tetrahedral field,* then, *it is the t_{2g}-orbitals which are most affected by the field.* Their energy level is raised whilst that of the e_g-orbitals is lowered.

The resulting energy difference between the t_{2g} and e_g orbitals, for a tetrahedral field, is summarised in Fig. 107b. The difference, Δ, is less than that for an octahedral field. This is because there are only four ligands in the tetrahedral field as compared with six in the octahedral field. Moreover, there is a ligand along each axis on the octahedral field and no ligand lies directly along any axis in the tetrahedral field. For ligands at the same distances from the central ion, it can be calculated that the purely electrostatic effect of a tetrahedral field will be approximately $\frac{4}{9}$ths that of an octahedral field.

On this basis, the t_{2g}-orbitals are raised in energy by $0·18\varDelta$ whilst the e_g-orbitals are lowered by $0·27\varDelta$.

(c) *Square planar field.* Four ligands in a square, planar arrangement around a central ion (Fig. 108) would have the greatest influence on a

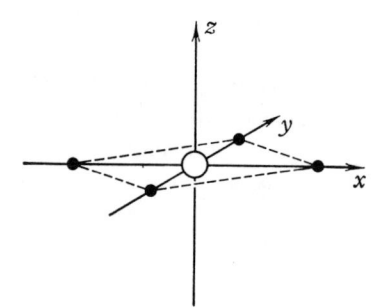

Fig. 108. Square planar arrangement of four ligands around a central ion. The effect of the ligand field is greatest on the $d_{x^2-y^2}$ orbital

$d_{x^2-y^2}$-orbital, so that the energy of this orbital will be raised most. The d_{xy}-orbital, lying in the same plane as, but between, the ligands will also have a greater energy though the effect will be less than on the $d_{x^2-y^2}$-orbital.

It is not possible to predict the exact effect of the ligand field on the remaining orbitals, though it will be less than on the $d_{x^2-y^2}$- and d_{xy}-orbitals, but the d_{yz} and d_{xz} pair will always be affected equally and will, therefore, remain degenerate.

A likely splitting of the d-orbitals in a square planar field is shown in Fig. 107c and a quantitative estimate of the splitting is given on page 192.

4 The filling of d-orbitals If less than ten electrons are available for occupying the five d-orbitals some measure of choice is open when the ligand field is strong enough to split the orbitals into different energy levels. For, if the energy difference is significant, electrons will fully occupy the orbitals of lower energy before entering those of higher energy.

(a) *Octahedral field*. Fig. 109 shows the way in which the *d*-orbitals will be occupied by electrons in an octahedral ligand field. *In a weak field*, the five *d*-orbitals will be occupied singly before any pairing of electrons takes place, as in the free atom (page 28). This is because the energy difference between t_{2g}- and e_g-orbitals will be too small to affect the issue. The fourth electron can either be placed in the e_g-orbital of higher energy or it can be paired with one of the t_{2g}-electrons. Both processes will require energy. The one because an orbital of higher energy has to be occupied; the other because elec-

No. of *d* electrons	Weak field High spin		Strong field Low spin	
	t_{2g}	e_g	t_{2g}	e_g
1	↓		↓	
2	↓ ↓		↓ ↓	
3	↓ ↓ ↓		↓ ↓ ↓	
4	↓ ↓ ↓	↓	↓↑ ↓ ↓	
5	↓ ↓ ↓	↓ ↓	↓↑ ↓↑ ↓	
6	↓↑ ↓ ↓	↓ ↓	↓↑ ↓↑ ↓↑	
7	↓↑ ↓↑ ↓	↓ ↓	↓↑ ↓↑ ↓↑	↓
8	↓↑ ↓↑ ↓↑	↓ ↓	↓↑ ↓↑ ↓↑	↓ ↓
9	↓↑ ↓↑ ↓↑ ↓↑	↓	↓↑ ↓↑ ↓↑ ↓↑	↓

Fig. 109. *The filling of d orbitals under the influence of weak and strong ligand fields, with the ligands arranged octahedrally*

trons repel each other. But in a weak field, it requires less energy to feed the fourth electron into an e_g-orbital than to pair it with one the t_{2g}-orbitals.

In a strong ligand field, with a larger difference in energy between the e_g- and t_{2g}-orbitals, electron pairing will begin when four electrons are available, the fourth electron pairing with a t_{2g}-electron instead of occupying a higher-energy e_g-orbital.

There is, therefore, a significant difference in the arrangement of *d*-electrons in a weak and a strong field, so far as ions with 4, 5, 6 or 7 *d*-electrons are concerned. Such ions are conveniently labelled d^4, d^5, d^6 and d^7 ions.

The number of unpaired electrons is greater in a weak field than in a strong field, for the strong field forces electrons to pair. Complexes with a weak ligand field are called *high-spin complexes*; those with strong ligand fields are called *low-spin complexes*. Alternatively, the terms low-field and high-field complexes are used.

The FeF_6^{3-} ion is an example of a high-spin complex whilst $Fe(CN)_6^{3-}$ is a low-spin complex, the field strength of CN^- ligands being much stronger than that of F^- ligands (page 189).

In a free Fe atom and a free Fe^{3+} ion, the electrons are arranged as follows:

	1s	2s	2p	3s	3p	3d	4s
Fe atom	2	2	6	2	6	↑↓ ↑ ↑ ↑ ↑	2
Fe^{3+} ion	2	2	6	2	6	↑ ↑ ↑ ↑ ↑	

In an octahedral field, however, the 3d-orbitals split into t_{2g}- and e_g-orbitals in the FeF_6^{3-} ion. The energy difference between the two sets of orbitals is not great enough to cause any spin-pairing, so that the arrangement of the five 3d-electrons is as shown in Fig. 110b. There

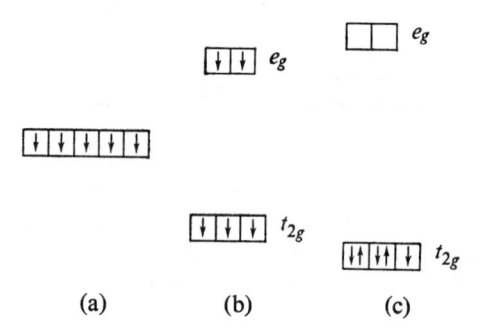

$$\boxed{↑↑}\ e_g \qquad \boxed{}\ e_g$$

$$\boxed{↑↑↑↑↑}$$

$$\boxed{↑↑↑}\ t_{2g} \qquad \boxed{↑↓↑↓↑}\ t_{2g}$$

(a)   (b)   (c)

Fig. 110. *Arrangement of d electrons in Fe^{3+} ion under different ligand fields. (a) Free ion. (b) In weak octahedral field, e.g. FeF_6^{3-}. (c) In strong octahedral field, e.g. $Fe(CN)_6^{3-}$.*

are five unpaired electrons so that a magnetic moment of 5·92 would be expected; the measured value is 6·0.

In the $Fe(CN)_6^{3-}$ ion the stronger field of the CN^- ions causes a greater energy difference between the t_{2g} and e_g-orbitals. As a result, spin-pairing occurs so that the five d-electrons are arranged as shown in Fig. 110c. There is only one unpaired electron and the expected magnetic moment would be 1·73; the measured value is 2·3.

It will be seen (page 166) that the high-spin complex, with no pairing of electrons corresponds with the ionic bonding in FeF_6^{3-} suggested by Pauling, whilst the low-spin complex, with full electron pairing, corresponds with the covalent bonding in $Fe(CN)_6^{3-}$.

(*b*) *Tetrahedral field.* The occupation of *d*-orbitals in a central ion surrounded by tetrahedral ligands exerting weak and strong fields is shown in Fig. 111. It will be seen that high- and low-spin complexes might be formed by d^3, d^4 d^5 and d^6 ions, but, in fact, low-spin complexes are extremely rare, and, possibly, non-existent. $ReCl_4^-$ is

No. of d electrons	Weak field High spin					Strong field Low spin				
	e_g		t_{2g}			e_g		t_{2g}		
1	↓					↓				
2	↓	↓				↓	↓			
3	↓	↓	↓			↓↑	↓			
4	↓	↓	↓	↓		↓↑	↓↑			
5	↓	↓	↓	↓	↓	↓↑	↓↑	↓		
6	↓↑	↓	↓	↓	↓	↓↑	↓↑	↓	↓	
7	↓↑	↓↑	↓	↓	↓	↓↑	↓↑	↓	↓	↓
8	↓↑	↓↑	↓↑	↓	↓	↓↑	↓↑	↓↑	↓	↓
9	↓↑	↓↑	↓↑	↓↑	↓	↓↑	↓↑	↓↑	↓↑	↓

Fig. 111. The occupation of d orbitals in a central atom surrounded by tetrahedral ligands exerting weak and strong fields

diamagnetic, which suggests no unpaired electrons, but the geometrical shape of the complex has not been established.

5 Measurement of Δ The splitting of *d*-orbitals by a ligand field into two sets with different energies provides a possible electronic energy change within a complex. An electron in an orbital of lower energy might be excited into an orbital of higher energy by the absorption of radiation. It so happens that the radiation necessary for such excitation lies in the visible region, and this explains why many complexes are coloured (page 190).

Study of the absorption spectra of complexes enables Δ values to be obtained. A band, or bands, in the absorption spectra can be related to the excitation of an electron, or electrons, between d-orbitals split by a ligand field. The ability to interpret such spectroscopic evidence is one of the advantages of crystal field theory.

The situation in the $Ti(H_2O)_6^{3+}$ ion is particularly simple, for there is only one d-electron involved. In the ground state, this will occupy one of the t_{2g}-orbitals, but absorption of radiation will excite it into one of the e_g-orbitals. The absorption spectrum of $Ti(H_2O)_6^{3+}$ shows a strong band corresponding to a wavelength of 0·5 μm. This represents an energy change of 239·1 kJ mol^{-1} (page 14) which must be the Δ value for this complex. Light of wavelength 0·5 μm is green. If such light is absorbed, the transmitted light appears purple (page 190) and the $Ti(H_2O)_6^{3+}$ ion is, in fact, purple.

With complexes containing more than one d-electron the possible energy changes between d-orbitals become more complex and the resulting absorption spectra contain more than one band. They can, nevertheless, often be interpreted to give Δ values. The absorption spectrum of $Ni(H_2O)_6^{2+}$ ions, for example, shows three strong absorption bands which can be accounted for if Δ is taken as being equal to 101·7 kJ mol^{-1}.

6 Factors affecting Δ   The actual value of Δ in any complex depends on the geometrical shape, the nature of the central ion and the nature of the ligands.

(*a*) *Effect of geometrical shape.* Figure 107 shows that the value of Δ for a tetrahedral complex is smaller than that for an octahedral complex with the same ligands at the same distance from the central ion, and the reason for this is explained on page 184.

Absorption spectra studies have been made on octahedral complexes much more extensively than on tetrahedral ones.

(*b*) *Effect of central ion.* For a given ligand, and given geometry, the Δ value depends on the valency of the central ion and on whether it is from an atom in the first, second or third transition series. The Δ values for ions within the same transition series, having the same valency, do not vary greatly.

For divalent ions from the first transition series, the Δ value is approximately 125 kJ mol^{-1}. The corresponding value for trivalent ions is approximately 250 kJ mol^{-1}. This effect is probably due to the fact that central ions with higher charge will polarise the ligands more effectively and, in this way, increase their ligand field.

For ions exerting the same valency the Δ value rises by about 30 per cent in passing from the first transition series, where 3*d*-electrons are involved, to the second, where 4*d*-electrons are involved. There is a similar rise in passing from the second to the third series. This effect is probably due to the fact that the *d*-electrons move further away from the nucleus in passing from a lower to a higher transition series.

(*c*) *Effect of ligands. The spectrochemical series.* For a given central ion and a given geometry the actual ligand has a marked effect on Δ. A ligand exerting a strong field will give a high Δ value; a weak ligand field gives a low Δ.

It is possible to list the common ligands in order of their field strength, the resulting list, as shown, being called the spectrochemical series:

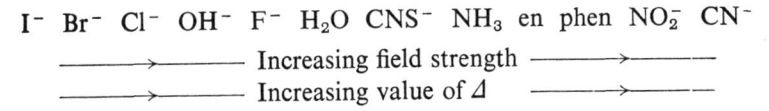

I^- Br^- Cl^- OH^- F^- H_2O CNS^- NH_3 en phen NO_2^- CN^-

—————→—————— Increasing field strength ——————→————

—————→—————— Increasing value of Δ ——————→————

Such an *order* applies more or less accurately no matter what central ion or what geometrical shape is concerned, though the absolute values of Δ may change considerably.

There are some obvious peculiarities in this order. It seems odd, for instance, that the charged OH^- ion should exert a smaller field than the uncharged H_2O molecule, or that the CN^- ion should exert a stronger field than the F^-. Such discrepancies can only be resolved by applying the more detailed ideas of ligand field theory (pages 198–206).

17 Some Applications of Crystal Field Theory

1 Effect of ligand field on colour Most complexes are coloured because they absorb light in the visible region. This is because the energy of light in the visible region is of just the right value to promote d-electrons from lower to higher energy levels, i.e. from t_{2g} to e_g in an octahedral complex. The energy required for such a promotion, Δ (page 183), will vary according to the ligand so that the colour of a complex is dependent on the ligands it contains. Ti(H$_2$O)$^{3+}$ which absorbs light of wavelength 0·5 μm is, for example, purple (page 188).

For a given central ion, replacement of the ligands by others further to the right in the spectro-chemical series (page 189) will give larger Δ values. This will lead to a lowering of the wavelength of the absorbed light, and a consequent shift from red towards blue in the colour of the absorbed light. The colour of the transmitted light will be complementary to that which is absorbed, as shown in the following summary:

Wavelength of absorbed light/μm	Wave-number of absorbed light/cm^{-1}	Colour of absorbed light	Colour of transmitted light
0·400–0·435	25,000–22,990	Violet	Yellow-green
0·435–0·480	22,990–20,830	Blue	Yellow
0·480–0·490	20,830–20,410	Green-blue	Orange
0·490–0·500	20,410–20,000	Blue-green	Red
0·500–0·560	20,000–17,800	Green	Purple
0·560–0·580	17,800–17,240	Yellow-green	Violet
0·580–0·595	17,240–16,810	Yellow	Blue
0·595–0·605	16,810–16,530	Orange	Green-blue
0·605–0·750	16,530–13,330	Red	Blue-green

The colour of anhydrous copper(II) sulphate(VI) is white because it absorbs light in the infra-red region and, therefore, reflects all visible wavelengths. The absorption is in the infra-red because the ligand field effect of sulphate(VI) ions is very small. Hydrated copper (II) sulphate(VI), containing Cu(H$_2$O)$_4^{2+}$ ions, is blue because it absorbs yellow light. Tetramminecopper(II) sulphate(VI), containing Cu(NH$_3$)$_4^{2+}$ ions, is violet because it absorbs yellow-green light. The Cu(en)$_2^{2+}$ ion is still deeper in colour, but the Cu(CN)$_4^{2-}$ ion, which absorbs light in the ultra-violet, is colourless.

Similar series of colour changes depending on the nature of the ligand are given below, for some chromium(III) complexes:

$Cr(NH_3)_6^{3+}$	$Cr(NH_3)_5(H_2O)^{3+}$	$Cr(NH_3)_4(H_2O)_2^{3+}$	$Cr(NH_3)_3(H_2O)_3^{3+}$
Yellow	Orange-yellow	Orange-red	Bright red
$Cr(NH_3)_2(H_2O)_4^{3+}$	$Cr(H_2O)_6^{3+}$	CrF_6^{3-}	$CrCl_6^{3-}$
Violet-red	Violet	Green-blue	Blue-green

and for some cobalt (III) complexes:

$Co(NH_3)_6^{3+}$	$Co(NH_3)_5Cl^{2+}$	$Co(NH_3)_4Cl_2^+$	$Co(NH_3)_3Cl_3$
Orange	Purple	Violet	Blue-green

The examples chosen probably over-simplify and exaggerate the effect for it is not always possible to account for colour changes so fully, and effects other than d-electron shifts may enter in.

2 Effect of crystal field on heats of hydration The heats of hydration of divalent ions between Ca^{2+} and Zn^{2+} do not rise steadily, as might be expected, but show maxima at V^{2+} and Ni^{2+} and a minimum at Mn^{2+} (page 169). Similarly the heats of hydration of the trivalent ions between Sc^{3+} and Ga^{3+} show a minimum at Fe^{3+} and a maximum at Cr^{3+}.

The hydrates concerned are all octahedral, high-spin complexes and the unexpected minima and maxima can be accounted for using the conception of *crystal field stabilisation energy* or ligand field stabilisation. This, in fact, was one of the early quantitative succesess of the theory.

In an octahedral field, the energy level of the e_g-orbitals is raised by $\frac{3}{5}\Delta$ whilst that of the t_{2g}-orbitals is lowered by $\frac{2}{5}\Delta$ (Fig. 104, page 181). For an octahedral complex with x-electrons in the higher energy orbitals and y-electrons in the lower energy ones, the energy difference, as compared with the energy level without any crystal field splitting, will be given by $(3x/5)\Delta - (2y/5)\Delta$. This quantity, with the sign reversed, is known as the crystal field stabilisation energy, which is sometimes abbreviated to CFSE.

The theoretical values for both high- and low-spin octahedral complexes, summarised below, are easily calculated from the arrangements of electrons given on page 185.

No. of electrons	1	2	3	4	5	6	7	8	9
High-spin	0·4Δ	0·8Δ	1·2Δ	0·6Δ	0	0·4Δ	0·8Δ	1·2Δ	0·6Δ
Low-spin	0·4Δ	0·8Δ	1·2Δ	1·6Δ	2·0Δ	2·4Δ	1·8Δ	1·2Δ	0·6Δ

For high-spin complexes it will be seen that the crystal field stabilisation energy is at a minimum (zero) for d^5 ions, such as Mn^{2+} and Fe^{3+}, and at a maximum for d^3 and d^8 ions such as V^{2+}, Ni^{2+}

and Cr^{3+}. This is why the measured heats of hydration show minima at Mn^{2+} and Fe^{3+}, and maxima at V^{2+}, Ni^{2+} and Cr^{3+}. If, in fact, the measured heats are corrected for each ion, by subtracting the calculated field stabilisation energy, the corrected heats of hydration lie on a smooth curve as shown by the dotted lines in Figs. 102 and 103 (page 170).

Similar results have also been obtained with the heats of formation of complexes, e.g. halides, other than hydrates.

3 CFSE values for octahedral, tetrahedral and square complexes

The crystal field stabilisation energies are very varied for different structures and for different ions. They have already been summarised for octahedral complexes on page 191. For tetrahedral complexes, the energy difference between e_g- and t_{2g}-orbitals is approximately $\frac{4}{9}$ths that of the corresponding energy difference, Δ, in an octahedral complex. The energy level of the t_{2g}-orbitals is raised by 0.18Δ whilst that of the e_g is lowered by 0.27Δ (page 184).

For a tetrahedral complex with x-electrons in the higher t_{2g}-orbitals and y in the lower e_g-orbitals, the increase in energy will be $0.18x\Delta - 0.27y\Delta$. This quantity, with the sign reversed, is the crystal field stabilisation energy (CFSE) for a tetrahedral complex in terms of the Δ value for an octahedral complex. The arrangement of electrons in the tetrahedral complex for both high- and low-spins is shown on page 187, so that it is a simple matter to work out the CFSE values, as is done below.

For a square complex the precise energy levels of the different d-orbitals have to be estimated. In what follows the energy lowerings for d_{xz}-, d_{yz}- and d_{z^2}-orbitals in a square field are taken as 0.51Δ, 0.51Δ and 0.43Δ respectively, whilst the energy rises for d_{xy}- and $d_{x^2-y^2}$-orbitals are taken as 0.23Δ and 1.22Δ respectively. The corresponding CFSE values are calculated from the electron distribution in a square field, using the same method as has been applied to the octahedral and tetrahedral fields.

The CFSE values obtained are collected together below in terms of Δ units.

	No. of electrons	0	1	2	3	4	5	6	7	8	9	10
High-spin	Octahedral	0	0.4	0.8	1.2	0.6	0	0.4	0.8	1.2	0.6	0
	Tetrahedral	0	0.27	0.54	0.36	0.18	0	0.27	0.54	0.36	0.18	0
	Square	0	0.51	1.02	1.45	1.22	0	0.51	1.02	1.45	1.22	0
Low-spin	Octahedral	0	0.4	0.8	1.2	1.6	2.0	2.4	1.8	1.2	0.6	0
	Tetrahedral	0	0.27	0.54	0.81	1.08	0.90	0.72	0.54	0.36	0.18	0
	Square	0	0.57	1.02	1.45	1.96	2.47	2.90	2.67	2.44	1.22	0

4 Use of CFSE values The following points illustrate the use that can be made of such CFSE figures.

(a) Occurrence of square complexes. Although tetrahedral high-spin complexes are known for all the divalent ions of the first transition series, the four-co-ordinated low-spin complexes which are known are all square. This is because the CFSE for the square, low-spin complexes is greater than that for the tetrahedral low-spin complexes and the difference in CFSE between the two structures is particularly high when Δ has a high value, i.e. for just those ligands which might be expected to give low-spin complexes.

Similarly, the high Δ values for elements in the second and third transition series means that square complexes are found here more commonly than in complexes from the first transition series. All the four-co-ordinated complexes of platinum and palladium, for example, are square and diamagnetic.

(b) Complexes of nickel(II). The four-co-ordinated complexes of nickel(II) were, at one time, regarded as either tetrahedral or square according to their magnetic moment, i.e. the number of unpaired electrons. Structure (i) for Ni^{2+}, enabling sp^3 hybridisation and with two unpaired electrons, could give tetrahedral, paramagnetic complexes, whereas structure (ii), enabling dsp^2 hybridisation and with no unpaired electrons, could give square complexes;

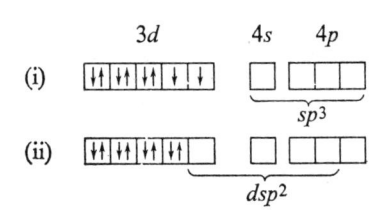

It is now thought that the majority of nickel(II) complexes are square and that nickel(II) will only form tetrahedral complexes in very unusual circumstances where steric or other effects might enforce a tetrahedral structure. The square complexes of nickel(II) are formed because the CFSE is particularly high for d^8 ions.

If the CFSE value drops because of a ligand with low Δ is used, nickel(II) complexes seem to polymerise into an octahedral arrangement rather than form tetrahedral structures.

(c) The structure of spinels. The spinels are a group of complex oxides

with a general formula $M^{2+}M_2^{3+}O_4$, in which M may be Mn, Fe or Co, or a number of other transitional or non-transitional metals. The common oxides, Fe_3O_4 and Mn_3O_4, are both spinels, with formula $Fe^{2+}Fe_2^{3+}O_4$ and $Mn^{2+}Mn_2^{3+}O_4$.

In the crystal structure of a spinel, one-third of the metallic ions are surrounded, tetrahedrally, by four O^{2-} ions, whilst two-thirds are surrounded, octahedrally, by six O^{2-} ions. In a *normal spinel*, all the M^{3+} ions occur at octahedral sites, and all the M^{2+} ions at tetrahedral ones. There is, however, an alternative structure, known as an *inverted spinel* structure, in which half the M^{3+} ions occupy tetrahedral sites whilst the other M^{3+} and M^{2+} ions occupy octahedral sites. It is found, too, that Mn_3O_4 has a normal spinel structure, whereas Fe_3O_4 has an inverted spinel structure.

This is due to the fact that, in a weak-field, such as that of oxide ions, Mn^{3+} (d^4) and Fe^{2+} (d^6) ions are both stabilised by CFSE in both octahedral and tetrahedral sites, whereas Mn^{2+} (d^5) and Fe^{3+} (d^5) ions are not. For Mn^{3+} and Fe^{2+} ions, the CFSE is greater for octahedral than for tetrahedral sites. In Mn_3O_4, therefore, the stablest arrangement is achieved when Mn^{3+} ions occupy all the octahedral sites as in a normal spinel. In Fe_3O_4, the stablest arrangement occurs when all the Fe^{2+} occupy octahedral sites as in the inverted spinel structure.

5 Effect of ligand field on bond distance Octahedrally arranged ligands affect the two e_g-orbitals, which are directed along the x-, y- and z-axes, more strongly than the t_{2g}-orbitals, which are directed between the axes. Equally, electrons in the e_g-orbitals repel the ligands more effectively than electrons in the t_{2g}-orbitals. This repulsion causes an increase in bond distance between central ion and ligand as the e_g-orbitals fill up.

This accounts for the measured bond distances between the central ion and oxygen in a series of oxides with sodium chloride structures, or similar compounds (Fig. 112). In the series from calcium oxide to zinc oxide the number of electrons in the various orbitals is as follows:

	Ca^{2+}	Sc^{2+}	Ti^{2+}	V^{2+}	Cr^{2+}	Mn^{2+}
No. of t_{2g}-electrons	0	1	2	3	3	3
No. of e_g-electrons	0	0	0	0	1	2

	Fe^{2+}	Co^{2+}	Ni^{2+}	Cu^{2+}	Zn^{2+}
No. of t_{2g}-electrons	4	5	6	6	6
No. of e_g-electrons	2	2	2	3	4

Addition of an extra t_{2g}-electron results in a lowering of the metal-oxygen bond distance, but addition of an e_g-electron causes it to rise. Similar results are obtained in a study of the metal-fluorine distances in a series of difluorides with rutile structures.

In both cases it is possible to calculate the theoretical effect of the e_g-electrons on the bond distance and to apply a correction to the measured distances. When this is done the corrected distances lie on a smooth curve as shown by the dotted line in Fig. 112.

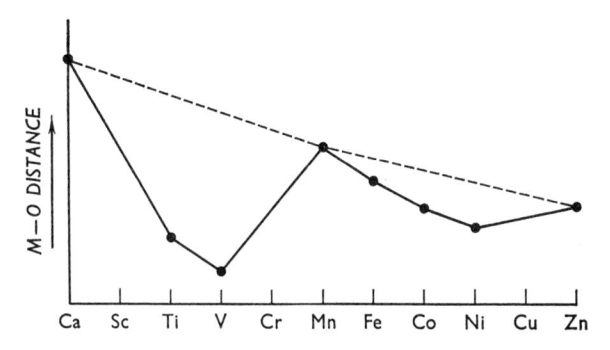

Fig. 112. *M–O distances in a series of comparable oxides*

6 Distorted octahedral complexes The regular shape of an octahedral complex is distorted if the d-orbitals of the central ion are not occupied symmetrically, because the symmetry of the electrical field is upset. This occurs whenever the t_{2g} or the e_g set of orbitals contains a number of electrons which will neither fill nor half-fill the set of orbitals. The effect of such a lack of symmetry on the geometrical shape of a complex is known as the *Jahn-Teller effect*. The effect applies to tetrahedral and to octahedral complexes, though only the latter will be considered here.

(*a*) *Asymmetric arrangement of e_g electrons.* There is no strong evidence to show that the asymmetric arrangement of electrons in t_{2g}-orbitals causes any distortion in the structure of a complex, but the effect of ligand fields is much greater on e_g-electrons (page 182) and if they are arranged asymmetrically distortion from the symmetrical octahedral structure will ensue.

The effect is of limited application, for it is only with d^4, d^7 and d^9 ions that the electrons can be arranged asymmetrically. For d^9 ions the asymmetry can occur in either strong or weak ligand fields; for d^7

ions strong fields are necessary; for d^4 ions, weak fields. The four possible asymmetric arrangements are shown below:

No. of d electrons	Weak ligand field High spin		Strong ligand field Low spin	
	t_{2g}	e_g	t_{2g}	e_g
4	↓ ↓ ↓	↓		
7			↓↑ ↓↑ ↓↑	↓
9	↓↑ ↓↑ ↓↑	↓↑ ↓	↓↑ ↓↑ ↓↑	↓↑ ↓

A pictorial idea of the cause of the distortion due to an asymmetric arrangement of electrons can be obtained by considering complexes of the Cu^{2+} (d^9) ion. The t_{2g}-orbitals are full; asymmetry is caused by incomplete filling of the e_g-orbitals. The three electrons in the e_g-orbitals may be distributed as $(d_{z^2})^1(d_{x^2-y^2})^2$ or as $(d_{z^2})^2(d_{x^2-y^2})^1$. The symmetrical distribution, $(d_{z^2})^2(d_{x^2-y^2})^2$, would give a regular

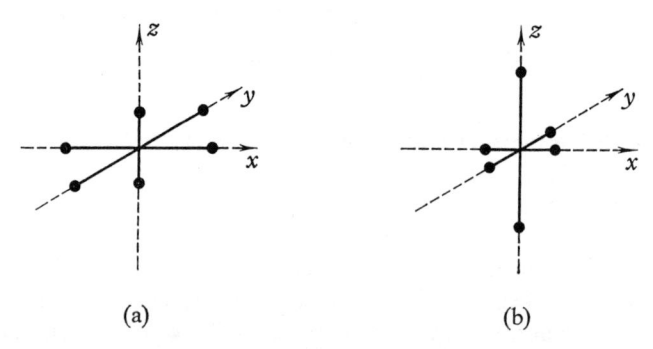

(a)   (b)

Fig. 113. Possible Jahn-Teller distortion in Cu^{2+} (d^9) complexes. In (a) there are four long and two short bonds, resulting from a $(d_{z^2})^1$ $(d_{x^2-y^2})^2$ distribution of electrons. In (b) there are two long and four short bonds, resulting from a $(d_{z^2})^2$ $(d_{x^2-y^2})$ distribution of electrons

octahedral complex, but the $(d_{z^2})^1(d_{x^2-y^2})^2$ distribution differs by having one less electron in the d_{z^2}-orbital. As this orbital is directed along the z-axis, the ligands on the z-axis would be drawn inwards if it only contains one electron. The resulting complex would, in this way, be slightly distorted, having two short bonds along the z-axis and four longer ones along the x- and y-axes (Fig. 113).

Alternatively, the electronic arrangement of $(d_{z^2})^2 d_{x^2-y^2})^1$ has the shortage of an electron in the $d_{x^2-y^2}$-orbital which is directed along the x- and y-axes. The distortion caused by such an arrangement of electrons results in a complex with four short bonds in the x- and y-directions with two longer bonds along the z-axis.

There is no theoretical way of deciding which of the two possible distortions is the more likely, but experimental results show that it is the arrangement of four short and two long bonds, i.e. the electronic arrangement $(d_{z^2})^2(d_{x^2-y^2})^1$ which predominates.

In the crystal of copper(II) chloride, for example, each Cu^{2+} ion is surrounded by six Cl^- ions, but four of them are at a distance of 0·23 nm whilst the other two are 0·295 nm away. Nickel(II) chloride has a similar crystal structure, but all the six Cl^- ions are equidistant from the central Ni^{2+} ion. This is because the Ni^{2+} ion, with eight d-electrons, has a symmetrical arrangement of electrons so that it forms perfectly symmetrical octahedral complexes.

Similarly, copper(II) fluoride has a distorted rutile structure (page 60) in which four F^- ions are 0·193 nm away from the Cu^{2+} ion whilst the other two F^- ions are 0·227 nm away. Chromium(II) fluoride is like copper(II) fluoride in having a similar distorted rutile structure. Like the Cu^{2+} ion, the Cr^{2+} ion, with four d-electrons, can have an asymmetrical arrangement of its electrons. The fluorides of Fe^{2+} (d^6), Co^{2+} (d^7), Ni^{2+} (d^8) and Zn^{2+} (d^{10}) all have regular rutile structures.

Many other similar applications of the Jahn-Teller effect are known.

(b) *Planar complexes as distorted octahedra.* If the bond lengthening along the z-axis becomes large enough what would originally have been regarded as an octahedral complex will become, in effect, a planar one. Some square planar complexes of Cu^{2+} can be regarded in this light, but the idea is much more fruitful as applied to d^8 ions such as Ni^{2+}, Pd^{2+}, Pt^{2+} and Au^{2+}.

With such a d^8 ion the expected arrangement of the significant electrons would be $(d_{z^2})^1(d_{x^2-y^2})^1$. This arrangement is symmetrical and the corresponding complexes would be perfectly octahedral, e.g. $Ni(H_2O)_6^{2+}$ and $NiCl_2$ (above). In a strong enough ligand field, however, the two electrons can be paired in the d_{z^2}-orbital, and, when this happens, the $d_{x^2-y^2}$-orbital is unoccupied. The corresponding repulsion for ligands along the z-axis and attraction for ligands along the x- and y-axes results in the formation of square planar complexes. A common example is provided by $Ni(CN)_4^{2-}$.

18 Ligand Field Theory

1 Introduction The crystal field theory topics discussed in the preceding chapters have taken no account of possible covalent bonding in complexes; the problems have been simplified by regarding them as purely electrostatic. But this is an over-simplification, for all the evidence† suggests that there is some measure of covalent bonding in complexes. As will be seen, this covalent bonding may consist of the formation of σ- and/or π-bonds between the central ion and the ligands.

Such bonding may be considered in terms of molecular orbitals formed by combination of atomic orbitals from the central ion and atomic orbitals from the ligands. Some of the molecular orbitals formed will be bonding orbitals; some will be anti-bonding. The procedure is very much the same as that adopted in dealing with bonding in simple molecules (page 92), but the problems with a complex are greater because the number of orbitals which have to be combined is larger. Moreover, the resulting molecular orbitals are de-localised (page 114); they are associated with the complex as a whole and not with any particular bond in the complex. This makes it more difficult to build up any pictorial ideas of bonding in a complex.

2 Combination of orbitals to form σ-bonds Atomic orbitals can only be combined within certain limitations of energy, overlap and symmetry as explained on page 85. To discover the molecular orbitals for a complex it is a matter of deciding which atomic orbitals of the central ion and ligands can be combined together. A full treatment uses the mathematical methods of *group theory*, but a more descriptive approach will be adopted here.

In an octahedral complex of an element from the first transition series, the orbitals on the central ion which must be considered, so far as σ-bonding (page 86) is concerned, are the $3d_{z^2}$ and $3d_{x^2-y^2}$ together with the $4s$, $4p_x$, $4p_y$ and $4p_z$. The $3d_{xy}$, $3d_{yz}$ and $3d_{xz}$, i.e. the t_{2g}, orbitals will not participate in any σ-bonding because they cannot effectively overlap with any orbital on the lines joining the ligands and the central ion. The t_{2g}-orbitals can, however, participate in π-bonding (page 202). Orbitals other than the $3d$, $4s$ and $4p$ are ruled out on energy grounds.

† The evidence rests mainly on physical measurements beyond the scope of this book. The methods of electron spin resonance, nuclear magnetic resonance and nuclear quadruple resonance are involved.

The six atomic orbitals of the central ion which can participate in σ-bonding combine with six ligand orbitals. For σ-bonding, these ligand orbitals must lie in the direction of the lines joining the ligand and the central ion. Such σ-orbitals, as they are called, will be s-orbitals, or some combination of s- and p-orbitals, pointing towards the central ion.

If the six ligands be denoted as L_1, L_2, L_3, L_4, L_5 and L_6 various possible arrangements of ligand orbitals in relation to orbitals on the central ion are shown, diagrammatically, in Fig. 114.

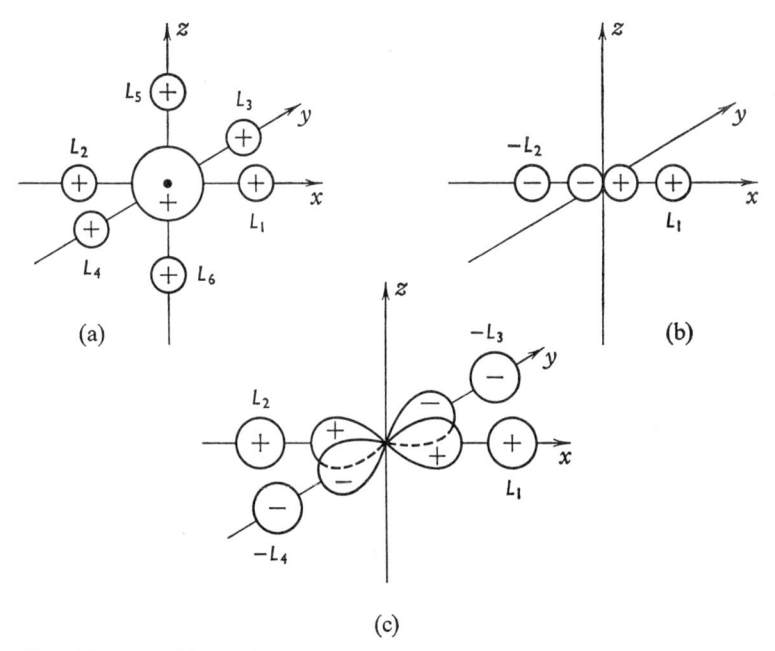

Fig. 114. Possible combinations of orbitals of the central ion and of the ligands. (a) s-orbital, (b) p_x-orbital, (c) $d_{x^2-y^2}$-orbital

The arrangement of ligand orbitals shown at (a) could be combined with the s-orbital of the central ion, for the symmetries are the same. The composite ligand orbital could be designated as $(L_1 + L_2 + L_3 + L_4 + L_5 + L_6)$. The composite ligand orbital which would be symmetrical with the p_x-orbital of the central ion would be $(L_1 - L_2)$, as shown at (b). Similarly, $(L_3 - L_4)$ and $(L_5 - L_6)$ would be symmetrical with the p_y- and p_z-orbitals, and $(L_1 + L_2 - L_3 - L_4)$ would be symmetrical with the $d_{x^2-y^2}$-orbital, as shown at (c).

In this way, the ligand and metal orbitals which can be combined together to give σ-bonds are summarised as follows:

Metal orbitals	Composite σ-orbitals for ligands
s	$(L_1+L_2+L_3+L_4+L_5+L_6)$
p_x	(L_1-L_2)
p_y	(L_3-L_4)
p_z	(L_5-L_6)
$d_{x^2-y^2}$	$(L_1+L_2-L_3-L_4)$
d_{z^2}	$(2L_5+2L_6-L_1-L_2-L_3-L_4)$

3 Bonding and anti-bonding orbitals   Two atomic orbitals combine together to give two molecular orbitals. One, of lowered energy, is a

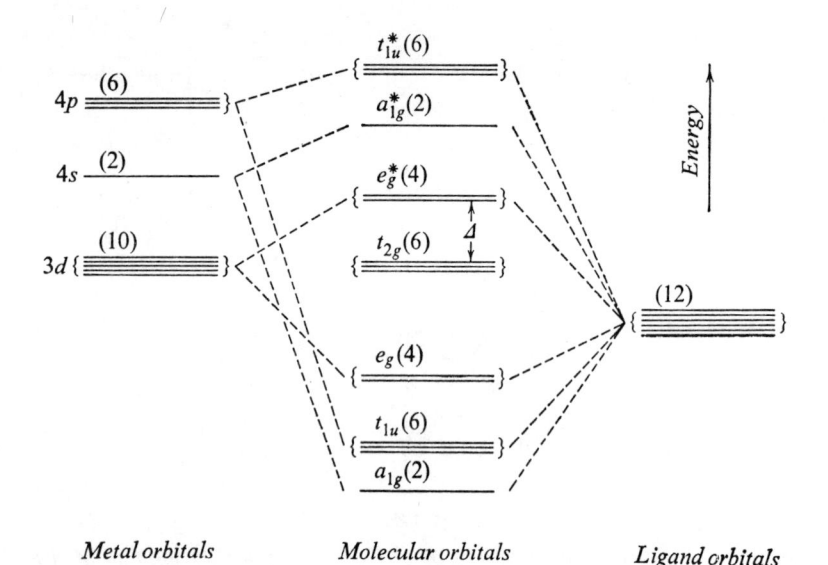

Fig. 115. *The combination of metal orbitals and ligand orbitals to form molecular orbitals in an octahedral complex*

bonding orbital, and the other, of higher energy, is an anti-bonding orbital (page 88).

The combination of the 12 orbitals given in the table at the end of Section 2 will, similarly, give 12 molecular orbitals, six of which will be bonding and six anti-bonding orbitals. The formation of these orbitals, and their relative energy levels, is shown schematically in Fig. 115. The actual energy levels have to be estimated in a semiempirical way, but the diagram given, and similar ones which can be drawn for tetrahedral and square complexes, are of fundamental importance in ligand field theory.

The metal orbitals are shown on the left-hand side. There are, in fact, nine of them but the three t_{2g} (d_{xy}, d_{yz} and d_{xz}) orbitals are unaffected by combination with ligand σ-orbitals. The six σ-orbitals of the ligands are shown on the right-hand side, and the 12 molecular orbitals, together with the three t_{2g}-orbitals, in the centre.

The metal s-orbital gives two molecular orbitals labelled as a_{1g}, and a_{1g}^*, the former being a bonding orbital and the latter an anti-bonding orbital. The three p-orbitals give three degenerate bonding orbitals, labelled t_{1u}, and three degenerate anti-bonding orbitals, labelled t_{1u}^*. The two e_g-orbitals give two degenerate bonding orbitals (e_g) and two degenerate anti-bonding orbitals (e_g^*). The three t_{2g} metal orbitals are unaffected; so far as σ-bonding is concerned they are *non-bonding* orbitals, though they do participate in π-bonding (page 202).

The nomenclature adopted is that of group theory. It is useful to remember that a-type orbitals are non-degenerate, e-type are doubly-degenerate and t-type are triply-degenerate. An asterisk is used to denote an anti-bonding orbital (page 83).

The number of electrons which each group of orbitals can hold is given in brackets in Fig. 115.

4 The filling of molecular orbitals Once the relative energy levels of the bonding and anti-bonding molecular orbitals in a complex have been established it is a simple matter to feed the available electrons in, in order of increasing energy.

With a d^1 complex, e.g. Ti(H$_2$O)$_6^{3+}$, there will be 13 electrons to be accommodated, one from the Ti^{3+} and two each from the six ligands. Twelve of these will occupy the bonding orbitals, a_{1g} (2), t_{1u} (6) and e_g (4). The other will be in a t_{2g}-orbital.

With d^n complexes, the t_{2g} and e_g^*-orbitals will fill up in just the same way as in crystal field theory (page 185). Whether high- or low-spin complexes will be formed will depend on the value of Δ, the energy difference between the t_{2g}- and e_g^*-orbitals (Fig. 115), and the Δ value will be essentially the same as in crystal field theory (page 181). The difference, on this point, between crystal field and ligand field theory is simply one of description. On crystal field theory ideas the t_{2g}- and e_g-orbitals have different energy levels simply because of the different effect on them of an electrostatic field. On ligand field theory, the t_{2g}- and e_g^*-orbitals have different energies because one set is anti-bonding and the other is nonbonding.

The conceptions of high- and low-spin complexes, their different magnetic properties, and the values of Δ are retained in the same form in ligand field theory as in crystal field theory. But ligand field theory maintains these ideas whilst at the same time admitting some measure of σ-bonding in a complex. Ligand field theory also admits some measure of π-bonding.

5 π-bonding in complexes The orbitals of the metal and the ligands which can be combined to form σ-bonds have been described in Section 2. It is also possible to combine orbitals between the metal and ligands so that π-bonds result.

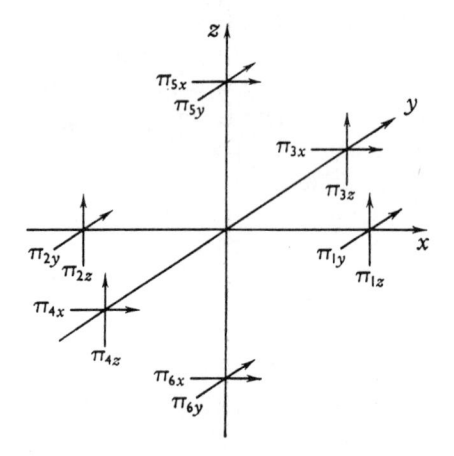

Fig. 116. Showing the axes of the ligand p-orbitals which can combine with orbitals on the central ion to form π-orbitals. The arrows point in the direction of the +lobe of the p orbitals

The metal orbitals available for this purpose are the p_x-, p_y-, p_z- and t_{2g}-(d_{xy}-, d_{yz} and d_{xz}-) orbitals. In simple language, they are the ones which can overlap with ligand orbitals 'sideways-on' to form π-bonds (page 85). These metal orbitals will combine with ligand π-orbitals of the right symmetry, i.e. the p-orbitals not directed along the line joining the ligand and the central ion (Fig. 116).

Two possibilities for combination of orbitals are shown schematically in Fig. 117. In (a), the d_{xz}-metal orbital can be combined with p-orbitals on the ligands if the composite p-orbital, known as a π-orbital, is made up of $(L_1p_z - L_2p_z + L_5p_x - L_6p_x)$ to get the symmetry right. Such a composite orbital is written as $(\pi_{1z}-\pi_{2z}+\pi_{5x}-\pi_{6x})$, π_{1z} being interpreted as a *p*-orbital for ligand number 1, with a nodal plane (page 84) in the *xy*-plane.

In Fig. 117 (*b*) a combination, $(L_1p_z+L_2p_z+L_3p_z+L_4p_z)$ or $(\pi_{1z}+\pi_{2z}+\pi_{3z}+\pi_{4z})$, will combine with a metal p_z-orbital.

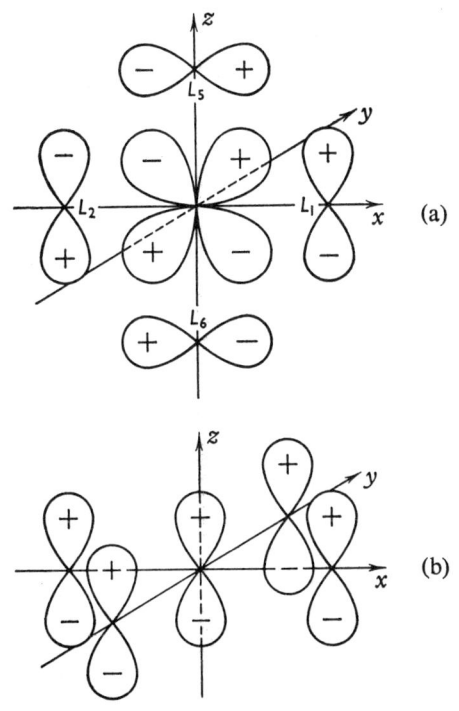

Fig. 117. (a) Combination of metal d_{xz}-orbitals with p_x- and p_z-orbitals of ligands. (b) Combination of metal p_z-orbital with ligand p_z-orbitals

On this basis, the orbitals which can be combined to give π-bonds are summarised as follows:

Metal orbitals	Composite π-orbitals for ligands
p_x	$(\pi_{3x}+\pi_{4x}+\pi_{5x}+\pi_{6x})$
p_y	$(\pi_{1y}+\pi_{2y}+\pi_{5y}+\pi_{6y})$
p_z	$(\pi_{1z}+\pi_{2z}+\pi_{3z}+\pi_{4z})$
d_{xy}	$(\pi_{1y}-\pi_{2y}+\pi_{3y}-\pi_{4y})$
d_{xz}	$(\pi_{1z}-\pi_{2z}+\pi_{5x}-\pi_{6x})$
d_{yz}	$(\pi_{3z}-\pi_{4z}+\pi_{5y}-\pi_{6y})$

Comparing this summary with that on page 200, it will be seen that the p_x, p_y and p_z metal orbitals can participate in both σ- and π-bonding.

6 Effect of π-bonding on complexes

The t_{2g}- (d_{xy}, d_{xz} and d_{yz}) orbitals do not participate in σ-bonding in complexes and, from that point of view, they are regarded as non-bonding orbitals (page 200). Because the e_g-orbitals *are* affected by σ-bonding there is an energy difference, Δ, between the t_{2g}- and e_g^*-orbitals.

The combination of t_{2g}-orbitals with ligand π-orbitals means, however, that the t_{2g}-orbitals are affected by π-bonding, and this affects their energy level, just as σ-bonding affected the energy level of the e_g-orbitals. Each of the three t_{2g}-orbitals forms two molecular orbitals, one of higher energy (anti-bonding) and one of lower (bonding).

This splitting of the t_{2g}-orbitals into two triply-degenerate sets affects the energy difference between the t_{2g}- and e_g^*-orbitals, i.e. the value of Δ. There are two cases.

(a) *Empty (acceptor) ligand orbitals of high energy.* If the ligand orbitals are of high enough energy, the splitting of the t_{2g}-orbitals will be as shown in Fig. 118a. If, also, the ligand orbitals are unoccupied, then the metal t_{2g}-electrons will go into the bonding molecular orbital of lower energy. The Δ value will, therefore, be increased to Δ'.

In other words, the ligand will exert a stronger field because of π-bonding. A ligand of this sort is referred to as an acceptor because of its empty π-orbitals. Phosphine and carbon monoxide are important examples (page 206).

This sort of π-bonding is sometimes referred to as metal-to-ligand (M $\rightarrow$ L) π-bonding. As it also tends to remove electronic charge away from the central metal onto the ligands it is also known as back donation or back bonding.

(b) *Occupied (donor) ligand orbitals of low energy.* The splitting of the t_{2g}-orbitals in this case is shown schematically in Fig. 118b. If the ligand orbitals are occupied, their electrons will occupy the lower, bonding t_{2g}-molecular orbital so that the metal t_{2g}-electrons will have to occupy the higher, anti-bonding orbital. Δ', in this case, will be lower than Δ, the ligand exerting a weaker field because of π-bonding.

Donor ligands include the halide ions and the water molecule. The π-bonds formed are known as ligand-to-metal ($L \rightarrow M$) π-bonds.

(c) *The spectrochemical series.* The introduction of the possibility of π-bonding in complexes throws more light on the order of ligands in the spectrochemical series. The original series, given on page 189, was considered from a purely electrostatic point of view. But if a ligand can form σ- and/or π-bonds with a central ion the field it will exert may be considerably affected.

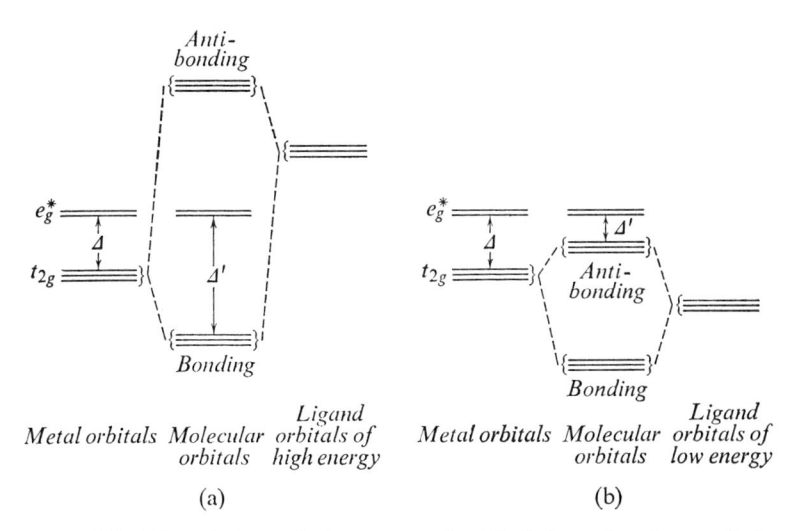

Fig. 118. *The splitting of the t_{2g} metal orbitals by π-bonding and the effect on the value of Δ*

It is, for instance, the acceptor nature of the CN^- ion in forming π-bonds which enables it to exert such a strong field effect and to head the spectro-chemical series. Conversely, it is the donor action of the halide ions which weakens their field, and the fact that the Δ value increases in the order $I^- < Br^- < Cl^- < F^-$ can be accounted for in terms of the smaller effect of I^- ions on the splitting of the t_{2g}-orbitals, as shown in Fig. 119.

Fig. 119. *I^- ions have a smaller effect on the splitting of the t_{2g} orbitals by π-bonding than F^- ions. As a result, the Δ value for I^- ions is less than that for F^- ions*

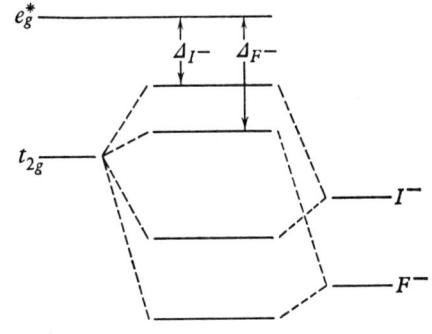

The spectro-chemical series passes, in fact, from I⁻, which is a strong π-donor, through weak π-donors and weak π-acceptors to CN⁻, which is a strong π-acceptor.

7 Examples of π-bonds in complexes The way in which the formation of π-bonds between metal and ligand can be used to explain known chemical facts can be illustrated by considering the structures of carbonyls and hydrocarbon complexes.

(a) Carbonyls. Carbon monoxide forms a number of complexes with transitional metal atoms, e.g. Ni(CO)$_4$, Fe(CO)$_5$ and Cr(CO)$_6$, in which the metal–carbon and carbon–oxygen bonds are colinear. At first sight it seems as though the bonding must be purely dative σ-bonding, for the carbon monoxide molecule has a lone pair on the carbon atom. Ni(CO)$_4$, for example, which is tetrahedral, might be regarded as having a structure shown in Fig. 120. Such a simple idea

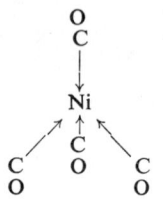

$$\begin{array}{c} O \\ C \\ \downarrow \\ Ni \\ \nearrow \uparrow \nwarrow \\ C \\ C \quad O \quad C \\ O \qquad\quad O \end{array}$$

Fig. 120. Over-simplified structure of Ni(CO)$_4$, *involving only ligand-to-metal σ-bonds*

does not, however, account for the facts. In the first place, carbon monoxide will not form stable complexes with elements other than transition elements. Nor, indeed, will it form dative bonds with such good acceptors as boron trifluoride, BF$_3$. This suggests that the *d*-electrons of a transitional metal must enter into the bonding in carbonyls. Secondly, the measured metal–carbon bond in carbonyls is shorter than would be expected for a single bond, i.e. some double-bonding is indicated. The carbon–oxygen bond is also lengthened in a carbonyl as compared with the distance in carbon monoxide itself, and a weakening of this bond in carbonyls is also shown by comparing the C—O absorption bands in the infra-red spectra of carbonyls and carbon monoxide.

Such facts can be accounted for if there is some double bonding between carbon and metal in carbonyls, and it is thought that this comes about by interaction between metal *d*-orbitals and the empty anti-bonding π-orbitals of the carbon monoxide molecule (page 98). Fig. 121 shows schematically how the orbitals could combine. This metal-to-ligand π-bonding is added to the ligand-to-metal σ-bonding.

Similar considerations apply in elucidating the structures of more complicated carbonyls, e.g. $Co_2(CO)_8$, $Mn_2(CO)_{10}$ and $Fe_3(CO)_{12}$, and in the complexes in which one or more CO ligands in a carbonyl are replaced by NO, nitrosyl, groups or by substituted phosphines, e.g. PCl_3 and $P(CH_3)_3$.

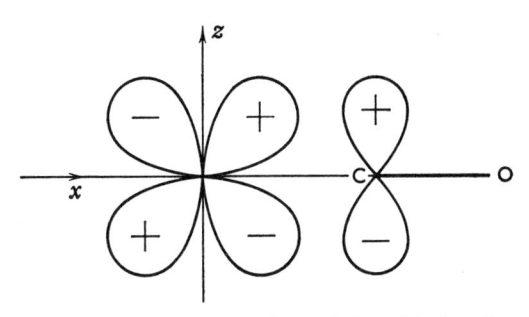

Fig. 121. Combination of metal d_{xz} orbital and p_z orbital of carbon monoxide

(b) *Hydrocarbon complexes.* In recent years many complexes between transitional metals and aromatic or unsaturated aliphatic hydrocarbons have been isolated and studied. Typical examples are shown below:

The molecules (*d*) and (*e*) are particularly unexpected. They are known as *sandwich molecules*, and a large group of sandwich compounds are now known for four-, five-, six-, seven- and eight-membered C_nH_n rings may form such compounds with many atoms, e.g. Ti, V, Cr, Mn, Co, and Ni.

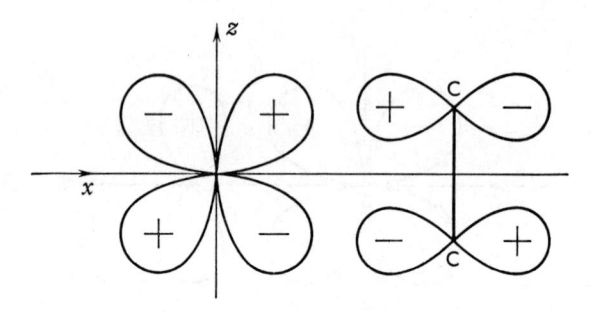

Fig. 122. Combination of metal d_{xz}-orbital with p_x-orbitals of ethene molecule

The bonding in all these hydrocarbon complexes is strongly influenced by π-bonding. Figure 122 shows, for example, the way in which a metal d_{xz}-orbital could form π-bonds with the p_x-orbitals on each carbon atom in an ethene molecule. Similar interaction between metal *d*-orbitals and the π-orbital systems of the ligands is used in explaining the structures of other hydrocarbon complexes.

19 Bonding in Metals

1 Introduction Metals have very distinctive properties. In particular, they are good electrical and thermal conductors; they are very opaque and have high reflecting power, i.e. they are lustrous; they are, generally, hard, ductile and malleable; they have high melting and boiling points, though the numerical range is wide; and they crystallise in crystals with high co-ordination numbers of 12 or 14 (page 223).

Such properties cannot be accounted for in terms of normal ionic or covalent bonding, so that the idea of a special metallic bond is necessary. The general picture of the state of affairs in the crystal of a metal is that positive ions of the metal are packed in a regular array but that the electrons liberated from the atoms of the metal can move through the crystal under the influence of an applied potential difference. It is, too, the relatively 'free' electrons that are responsible for the binding within the crystal. It is significant that the only atoms which are metallic are those with relatively low ionisation energies so that 'free' electrons can be made available.

The nature of the bonding in metals is peculiar, however, because there is such a shortage of electrons. Lithium, for example, has only three electrons per atom, and two of them, the $1s$-electrons, are very firmly held by the nucleus. In lithium, then, there is only one electron per atom available for any metallic bonding and yet each lithium has eight, or fourteen, near neighbours in its crystal (page 225). There is certainly no obvious way in which a lithium atom could form eight or fourteen bonds.

It is impossible, then, for anything like individual bonds to exist in metals. The few electrons available for bonding purposes must be completely delocalised (page 114). Molecular orbitals can be established for the metal as a whole, and this approach leads to the idea of energy bands within the crystal of a metal.

THE BAND MODEL

2 Energy bands The arrangement of electrons in a single sodium atom is as follows:

$1s$	$2s$	$2p$	$3s$
2	2	6	1

each electron being in a particular energy level. If two sodium atoms are brought close to each other the 3s-orbital of the one can combine with the 3s-orbital of the other to form two molecular orbitals as, for comparison, in the hydrogen molecule (page 92). One of the molecular orbitals will be of higher, and the other of lower, energy, as in Fig. 55, page 92. If a third sodium atom is introduced, three molecular orbitals will be possible; and for n atoms, n molecular orbitals can be formed. The building up of such molecular orbitals can be represented as in Fig. 123. The final number of molecular orbitals will be equal to the number of atoms concerned, and, as this will be

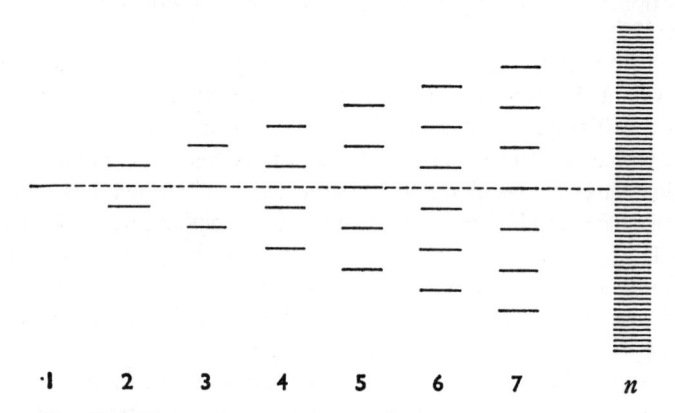

Fig. 123. *The combination of n atomic orbitals to form n molecular orbitals in an energy band*

very, very large for any actual crystal of sodium, the energy levels of the molecular orbitals will be so close that they can be regarded as forming an energy band.

In a similar way it might be possible to combine 2p-, 2s-, and 1s-orbitals to form energy bands in a crystal. The formation of an energy band will depend, however, not only on having enough atoms to interact but on the atoms being close enough together. In free sodium atoms, in the vapour state, there are no energy bands: the electrons occupy atomic orbitals with separate and distinct energy levels. As the sodium atoms are brought closer and closer together the interaction between the various atomic orbitals increases so that the 'width' of the resulting bands also increases.

Such band widening as the interatomic distance decreases is shown for the 2s, 2p, 3s and 3p levels in a sodium crystal in Fig. 124. It will be seen that the higher energy orbitals interact to form energy bands at distances greater than the lower energy orbitals. In an actual sodium crystal, with interatomic distance of 0·371 nm, the energy bands corresponding to the 3p and 3s levels are the significant ones and, as seen in Fig. 124, they overlap to some extent.

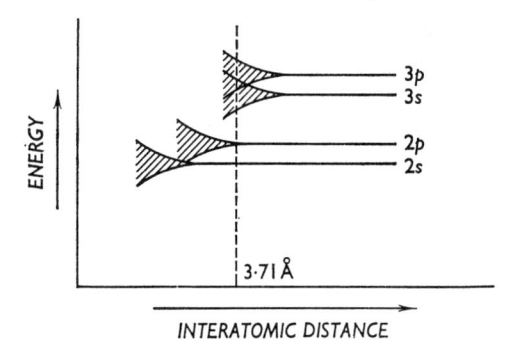

Fig. 124. The widening of energy bands as the interatomic distance is decreased, and the overlap of 3p and 3s bands in a sodium crystal

The situation can also be summarised, pictorially, as in Fig. 125. (*a*) shows the discrete energy levels (not to scale) of a free sodium atom; (*b*) shows the 3s and 3p bands developing as the sodium atoms

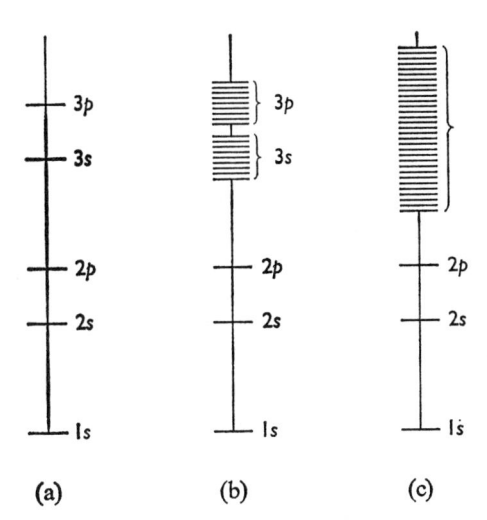

Fig. 125. The development of overlapping 3p and 3s energy bands in a crystal of sodium from the atomic levels in a free atom

come together; (c) shows the overlap of the 3s and 3p bands when the atoms attain the equilibrium position in the crystal.

The energy bands must be considered, of course, as belonging to the crystal as a whole; that is a measure of the complete delocalisation. It is the interaction of a large number of atomic orbitals of the right energy, the right symmetry and, in a crystal, sufficient overlap (page 85), which makes the formation of energy bands possible.

3 Soft X-ray spectra The existence of energy bands postulated in the preceding section is borne out by a study of soft X-ray spectra. It is, indeed, a study of such spectra which provides experimental information about the bands.

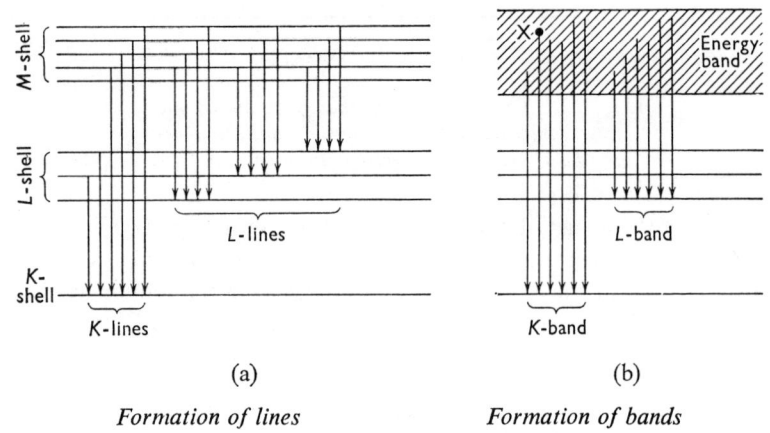

Fig. 126. *The formation of lines and of bands in soft X-ray spectra*

(a) *Formation of lines* (b) *Formation of bands*

When a cooled metal is bombarded by a beam of high-speed electrons some of the electrons in a low energy level of the metallic atom may be energised sufficiently to be ejected. Electrons taking the vacant place from higher levels will cause an emission of radiation in the X-ray region dependent on the energy changes involved. Electronic movements into the K-shell (principal quantum number = 1, page 15) give rise to K-lines in the X-ray spectrum; electrons moving into the L-shell give L-lines, and so on; as illustrated in Fig. 126a.

The resulting X-ray spectra provide extremely interesting results. If the metal involved is present as a vapour at low pressure a complete line spectrum results, but as the pressure is increased, i.e. as the atoms get closer together, the K- and L-lines merge into bands. Similar bands also occur when the metal is in its solid form. These bands in the spectrum show the presence of energy bands in the metal. Once the atoms of the metal get close enough together electrons in the energy band of the metallic atoms moving into the K- and L-shells give spectral bands rather than spectral lines, as indicated in Fig. 126b.

4 $n(E)$ curves The idea of energy bands, established both theoretically and experimentally, gives a new picture of the energy levels within the crystal of a metal. The inner electrons of the atoms occupy localised orbitals and not bands, but the valency orbitals and higher ones merge into energy bands representing a state of high delocalisation. It is the arrangement of the available electrons in the available energy levels which determines the characteristics of a metal, just as it is the arrangement of the electrons in the available energy levels of a free atom which determine the nature of the free atom (page 25).

Just as line spectra, too, throw light on the arrangement of electrons in free atoms (page 17) so do soft X-ray spectra help to elucidate the arrangement of electrons in the energy bands of metal crystals. It is the variation in intensity of the band in a soft X-ray spectrum that gives valuable information. An energy band covers a range of energies and the width of the range shows up in the width of the spectral band. Moreover, the way in which the electrons are distributed in the energy band between the various possible energy levels or states shows up in the spectral band. If, for instance, the majority of electrons in an energy band have an energy, X, then the spectral band in the K-spectrum will be most intense at the point corresponding to the energy change between X and the K-level (Fig. 126b).

So-called $n(E)$ curves are used to show the complete distribution of electrons between the range of energy levels in an energy band, $n(E)dE$ being the number of electrons per unit volume of material with energies between E and $(E+dE)$. $n(E)$ plotted against E gives the energy distribution of the electrons in a band.

Figure 127a shows a typical $n(E)$ curve for a metal at normal temperatures. The energy of the electrons in the band which the curve represents varies between A and B, and the energy difference between A and B measures the width of the energy band; it varies from metal to metal but is of the order 1–10 electron volt (page 14). The curve shows that C is the most favoured energy, and the area under the curve is proportional to the total number of electrons in the particular band.

As the temperature is decreased the right-hand side of an $n(E)$ curve becomes more and more vertical, until, at absolute zero, it becomes completely vertical (Fig. 127b). This means that the high-energy end (head) of an energy band is sharp at absolute zero but becomes more and more diffuse as the temperature rises.

The effect of temperature on the spectral band is due to the effect of temperature on the occupancy of the available energy levels or states within a band. At absolute zero, it is assumed that the available

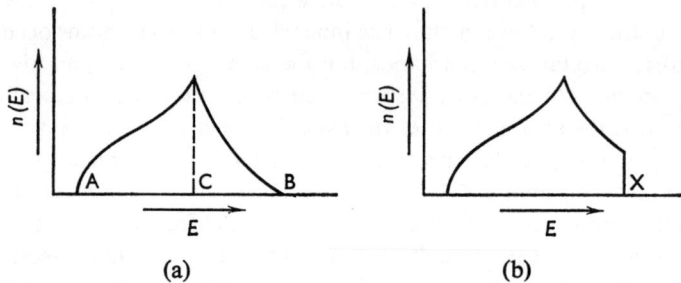

Fig. 127. Typical $n(E)$ curves. (a) *At normal temperature.* (b) *At absolute zero*

electrons occupy all the lowest levels within a band. If the highest of these occupied levels has an energy of X there will be no electrons with a higher energy value. The right-hand boundary of the $n(E)$ curve will, therefore, be at X. At higher temperatures thermal energy will promote at least some electrons into levels with energy higher

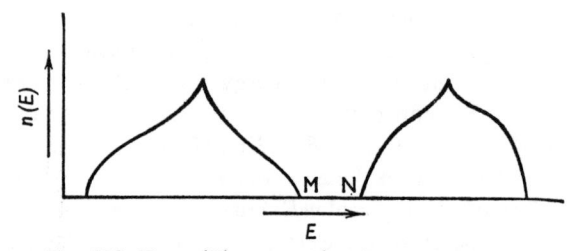

Fig. 128. Two $n(E)$ curves showing an energy gap

than X so that the vertical boundary of the $n(E)$ curve disappears. Fig. 127b merges into Fig. 127a.

Each band in a soft X-ray spectrum will have a corresponding $n(E)$ curve so that there may be more than one as shown in Fig. 128. In such a case it will be seen that there are energy gaps, i.e. certain energies between M and N for which there are no electrons. The gaps may be wide or narrow, or non-existent when two $n(E)$ curves overlap.

5 $N(E)$ curves The $n(E)$ curves refer to the existing distribution of electrons between the various possible energy levels or states in a band. The actual levels or states that exist in the band may differ because all the levels are not necessarily occupied by an electron, and an unoccupied level will not be indicated in the spectral band. To obtain a curve showing how the electron states are distributed over the energy range within a band it is necessary to introduce a new quantity $N(E)$ defined so that $N(E)dE$ gives the number of energy states with energy between E and $(E+dE)$.

At absolute zero the available electrons will be in the lowest energy levels so that n electrons will occupy the $n/2$ lowest energy states,

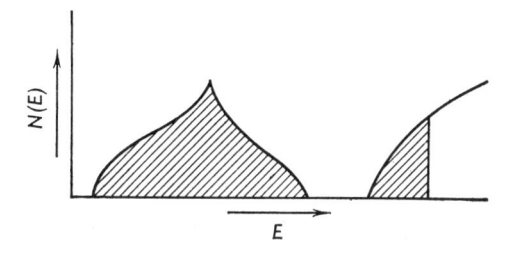

Fig. 129. Two $N(E)$ curves with an energy gap. The shaded portions show the energy states which would be occupied at absolute zero

each state containing two electrons with opposed spins. Because all these $n/2$ energy levels are full the $n(E)$ and $N(E)$ curves are of the same shape, though the vertical scale has to be adjusted to ensure that the area under the $n(E)$ curve, which gives the total number of electrons, is twice that under the $N(E)$ curve, which gives the total number of electron states (each holding two electrons).

At temperatures above absolute zero not all the available energy states in a band will be occupied so that the $N(E)$ curves will extend beyond the $n(E)$ curves. The two can be related, to some extent, by shading in the area under an $N(E)$ curve which represents the electron states that would be fully occupied at absolute zero. Figure 129 shows a typical $N(E)$ curve with the shaded portion representing the states which would be occupied at absolute zero. There are not enough electrons available to occupy all the available states. At absolute zero the available electrons will occupy the lowest states, but at higher temperatures the electrons will be distributed differently between the available states.

Like $n(E)$ curves, $N(E)$ curves may be separated by an energy gap, as in Fig. 129. Alternatively, two $N(E)$ curves may overlap, and, as will be seen (page 217), it is the shape of the $N(E)$ curves which is of vital importance in conductivity considerations.

6 Binding energy in metals The shortage of electrons for bond formation in metals, and the complete delocalisation of the bonds concerned, suggests that the actual binding forces in a metal might be weak. On the basis of individual bonding between two adjacent atoms the bonding in a metal crystal is, in fact, weaker than covalent bonding, but the very large number of bonds in the metal crystal make up, so far as total binding energy is concerned, for any weakness in the individual bonds.

The individual bond weakening, as compared with a covalent bond, is best illustrated in terms of interatomic distance. Diatomic molecules of the alkali metals, e.g. Na_2, exist in the vapour state and it seems likely that they contain single covalent bonds. The interatomic distances in these diatomic molecules are considerably smaller than the distances in the crystal of the metal, which suggests that the covalent bond is stronger than the bonding in the metallic crystal. The figures, in nm, are summarised below:

	Li	Na	K	Rb	Cs
In diatomic molecule	0·267	0·308	0·392	0·422	0·450
In crystal	0·304	0·372	0·462	0·486	0·524

The bonds can also be compared in terms of the bond energy of the covalent bond in the diatomic molecules and the heats of dissociation or sublimation of the metallic crystals. The bond energy gives a measure of the energy required to split the covalent bond. The heat of dissociation measures the energy necessary to break down the binding forces in the crystal of a metal to produce free atoms. The following figures show that the overall binding energy in the crystal of a metal is greater than in covalently bound diatomic molecules. The figures are given as kJ (gram-atom)$^{-1}$, which means that the covalent bond energies, which are normally expressed in kJ mol^{-1}, have been halved.

	Li	Na	K	Rb	Cs
In diatomic molecule	54·4	37·7	25·1	25·1	20·9
In crystal	134	96·2	79·5	75·3	66·9

The bonding forces in a metal crystal may, therefore, be weak if considered on an interatomic basis but, as they are widely spread throughout the whole crystal, the overall binding is strong.

7 Conductors and insulators Electrical conductivity is associated with partially filled energy bands or with filled energy bands overlapping empty or partially filled ones. There must be unoccupied energy states close to the occupied ones, so that electrons can easily be promoted.

In a metal crystal not under any applied potential difference equal numbers of electrons can be regarded as moving in all directions through the crystal; there is no resultant flow in any one direction. When a potential difference is applied, however, electrons moving in the direction of the applied field will have higher energy than those

Fig. 130. N(E) curve for univalent metal. The 3p and 3s bands overlap, and there are enough electrons to half fill the 3s band

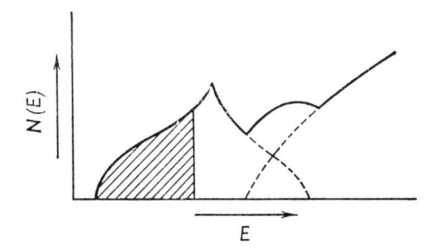

flowing in the opposite direction because their kinetic energy is increased. But electrons moving in the direction of the applied field will only be able to 'take-up' this extra energy if there is a nearby electron state of higher energy level to accommodate them, and there will be no such electron states in a completely occupied energy band.

In a sample of sodium, for example, containing n atoms there will be n valency electrons to be accommodated. The 3s band, by itself, would contain n levels (page 210) so that it could hold $2n$ electrons, two (with opposed spins) in each level. The overlap of the 3p and 3s bands provides an even greater number of levels so that sodium is a good conductor. The state of affairs is shown in Fig. 130. The 3p and 3s bands overlap, and the 3s band is shown as being half-full. This type of curve is typical for univalent metals.

Similarly, the overlapping 3p and 3s bands are not filled by the two valency electrons of a divalent metal, or by the three valency electrons of a trivalent metal. Magnesium and aluminium, for example, are conductors and the typical $N(E)$ curves for divalent and trivalent metals are shown in Figs. 131 and 132.

The conductivity does not rise in proportion to the number of valency electrons. In aluminium for example, the 3s band can be regarded as being full so that it will contribute nothing to the conductivity. The univalent metals, e.g. the alkali metals and copper,

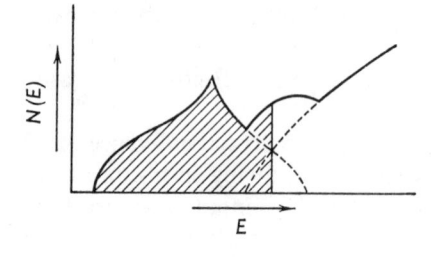

Fig. 131. $N(E)$ curve for a divalent metal with overlapping 3p and 3s bands. There are twice as many electrons available as in univalent metals (Fig. 130)

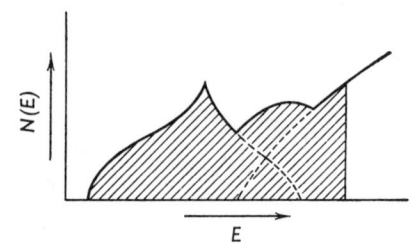

Fig. 132. $N(E)$ curves for trivalent metals with overlapping 3p and 3s bands. Compare Figs. 130 and 131

silver and gold, have the highest conductivities. They have the lowest number of valency electrons but there are a lot of unoccupied energy states as shown in Fig. 130.

In diamond, with four valency electrons, both the 3s and 3p bands would be full and the gap between these bands and the next higher

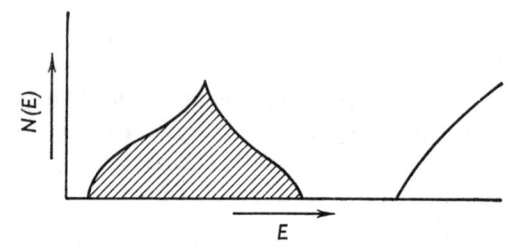

Fig. 133. A typical insulator with a filled band and a large gap between it and the next higher band

one is too big to be bridged by a small potential difference. Diamond is, therefore, an insulator, though very large potential differences will, of course, break down the insulation.

Figure 133 represents the state of affairs in a typical insulator.

8 Effect of temperature and impurities on conductivity Both rise in temperature and the presence of impurities tend to lower the conductivity of a metal, i.e. raise its resistance.

(*a*) *Effect of temperature.* It might be expected that a rise in temperature would increase the conductivity of a metal. In a simple way, the expansion of the metal might give more room for the electrons to move in, or the thermal energy might, perhaps, be expected to promote some electrons into partially occupied bands. In general, however, the conductivity of a normal metal decreases with rise in temperature.

This is because increased temperature produces increased thermal vibration which upsets the regularity within a crystal. The vibrating atoms cause a vibrating field which scatters the electrons into different energy states and lowers their freedom of motion. A quantitative treatment can, in fact, show that the scattering would be expected to be proportional to the absolute temperature which accounts for the proportionality between resistance and absolute temperature.

(*b*) *Effect of impurities.* The regular array and regular field within a crystal of a metal can also be disturbed by small quantities of impurities. As a result, the conductivity may be greatly decreased by small amounts of impurities. The following figures illustrate, for example, the effect of adding manganese to copper:

Per cent of Mn	0	1	2	3
Conductivity/Ω^{-1} cm^{-1}	$5 \cdot 9 \times 10^5$	$2 \cdot 0 \times 10^5$	$1 \cdot 2 \times 10^5$	$0 \cdot 8 \times 10^5$

9 Semi-conductors Whilst the effect of increased temperature and the presence of impurities is to lower the conductivity of normal metals there are some materials which may be insulators under normal conditions but which become conductors when the temperature is raised or when certain impurities are added. Such materials are known as semi-conductors, and the conductivity arises in three different ways.

(*a*) *Intrinsic semi-conductors.* Pure graphite, pure germanium and pure grey-tin are some of the relatively few examples of semi-conductors. In such materials the energy gap between the highest filled band and the next empty one is very small (Fig. 134). At absolute zero such materials would be insulators but an increase in temperature can produce enough thermal energy to promote some electrons from the filled band into the next higher one, across the small energy gap; a conductor results. What is more, the number of excited electrons increases as the temperature is increased so that the conductivity of intrinsic semi-conductors increases with rise of temperature.

(b) N-*type semi-conductors*. The addition of an impurity to a metal might provide additional energy levels, and, if these levels are correctly related to the bands within the pure metal, conductivity may result. If the impurity contains a full energy level just below that of an

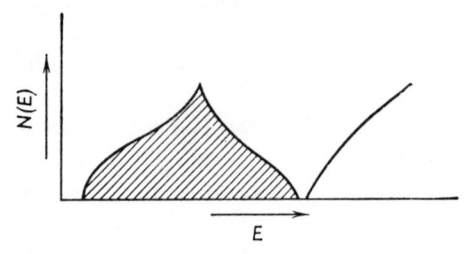

Fig. 134. An intrinsic semiconductor, with a very small gap between the full and empty bands

empty band in the pure metal, the electrons from the impurity might have enough energy to pass into the empty band in the metal (Fig. 135). This happens when arsenic or antimony are added to germanium. The passage of electrons from an energy level in the arsenic or antimony into one in the germanium causes the germanium to become negatively charged. It is, therefore, known as n-type germanium.

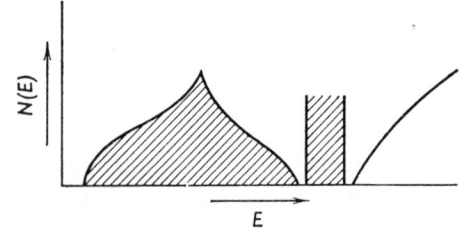

Fig. 135. An n-type semiconductor. The full band of impurity lies just below an empty band of the pure metal

The electrons in the impurity will have more energy and, therefore, more chance of passing into the higher band of the metal at higher temperatures. The conductivity of n-type semiconductors increases with temperature because of this. If only a small amount of impurity is present, however, there will be a limited number of electrons which can be promoted from the impurity so that the rise in conductivity with increasing temperature may be halted, and even reversed, if the temperature is increased sufficiently.

(c) P-*type semi-conductors*. If the impurity contains an empty energy level (Fig. 136) just above a full band in the pure metal, the electrons from the full band in the pure metal might be able to pass into the empty level of the impurity. This happens when gallium or indium are added to germanium. Passage of electrons from the germanium to the gallium or indium causes the germanium to become positively charged. It is called p-type germanium. The effect of temperature is the same as for n-type germanium.

Fig. 136. A p-type semi-conductor. The empty band of the impurity lies just above a full band of the pure metal

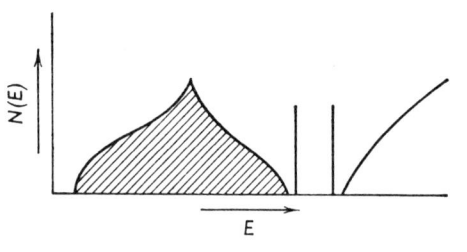

Semi-conductors are of great importance in making transistors.

STRUCTURES OF METALLIC CRYSTALS

10 Types of structure The units packed together in the crystal of a metal are all alike so that the radius ratio (page 55) must be 1. The resulting structures have high co-ordination numbers, and the great majority of metals crystallise in one of three structures:

(a) Cubic close-packed, or face-centred cubic, structure;

(b) Hexagonal close-packed structure; and

(c) Body-centred cubic structure.

Many metals are, in fact, polymorphous. Nickel and cobalt, for example, can have cubic, or hexagonal close-packed structures; iron can have a cubic close-packed, or a body-centred cubic, structure.

The incidence of the above structures amongst the metals is summarised in the table below:

Li	Be									
c	b									
Na	Mg									
c	b									
K	Ca	Sc	Ti	V	Cr	Mn	Fe	Co	Ni	Cu
c	a,b	a,b	b,c	c	b,c	—	a,c	a,b	a,b	a
Rb	Sr	Y	Zr	Nb	Mo	Tc	Ru	Rh	Pd	Ag
c	a	b	b,c	c	b,c	b	a,b	a	a	a
Cs	Ba	La	Hf	Ta	W	Re	Os	Ir	Pt	Au
c	c	a,b	b,c	c	c	b	a,b	a	a	a

Crystal structures of the true metals

The metals in the later B sub-groups, and manganese, have more complicated structures. The metals in the larger group, summarised above, with simple structures are sometimes known as *true* or *normal metals*.

11 Close-packed structures Both (*a*) and (*b*) in the preceding section are known as close-packed structures because they contain units (taken as being spherical) which are packed together in the tightest possible way to fill the maximum amount of available space.

In two dimensions, such packing can be achieved in only one way, as shown in Fig. 137a. The centres of adjacent spheres lie on the corners of equilateral triangles, and each sphere is in contact with six others. The spheres occupy 60·4 per cent of the available space. The only

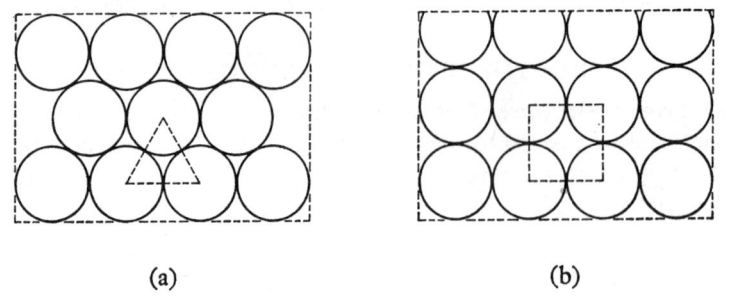

(a)   (b)

Fig. 137. Two possible ways of close-packing spheres in two dimensions. (a) is the most effective way

other way of packing spheres at all tightly together, shown in Fig. 137b, is less effective. Here the centres of the spheres lie on the corners of a square and each sphere is in contact with four others. The spheres occupy only 52·4 per cent of the available space.

A second layer of spheres fitted into half of the hollows in the layer shown in Fig. 137a gives the best close-packing for two layers. The centres of the spheres are at the corners of a regular tetrahedron (Fig. 96).

There are, now, two alternative ways of fitting in a third layer of spheres, each giving an equally close-packed structure. The spheres of the third layer must rest in half of the hollows of the second layer, but there are really two sets of hollows which differ in their relation to the first layer. Those hollows marked with a dot in Fig. 138 lie directly above the spheres in the first layer, whilst the hollows which are not marked lie above the holes in the first layer.

If the third layer of spheres is placed in the marked hollows, the third layer will be exactly the same as, and lie directly above, the first layer. It is convenient to label such an arrangement as ABA. If it is repeated, the arrangement of layers becomes ABABABAB..., i.e. the pattern repeats itself every third layer. This is the arrangement in the hexagonal close-packed structure (Fig. 139).

Fig. 138. Close-packing of two layers of spheres. The upper layer is shown by dotted lines, and there are two sets of hollows in this layer

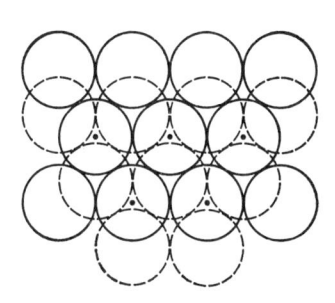

If, however, the third layer is placed so as to lie over the holes in the first layer the arrangement will be ABC. If this arrangement is repeated it becomes ABCABCABC..., i.e. the pattern repeats itself every fourth layer. This is the cubic close-packed structure (Fig. 140).

Fig. 139. Hexagonal close-packed structure showing a co-ordination number of 12

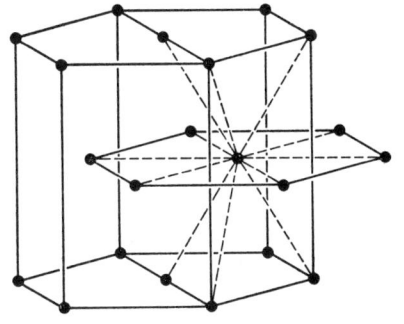

(a) *Hexagonal close-packed structure.* The arrangement of atoms in this structure is shown most conveniently as in Fig. 139. The three layers, ABA, can be clearly seen one above the other, and the 12 nearest neighbours to one atom are shown, giving a co-ordination number of 12.

(b) *Cubic close-packed or face-centred cubic structure.* Figure 140 shows the arrangement of atoms in this structure in two ways. On the left-hand side the four close-packed layers, ABCA, one above the other, are shown. At first sight it is not readily apparent that such a structure has cubic symmetry, but the same arrangement of atoms is shown on the right-hand side as a face centred cubic structure. The diagonal of the cube is normal to the planes in which the close-packed atoms lie, and, as there are four diagonals to the cube so there are

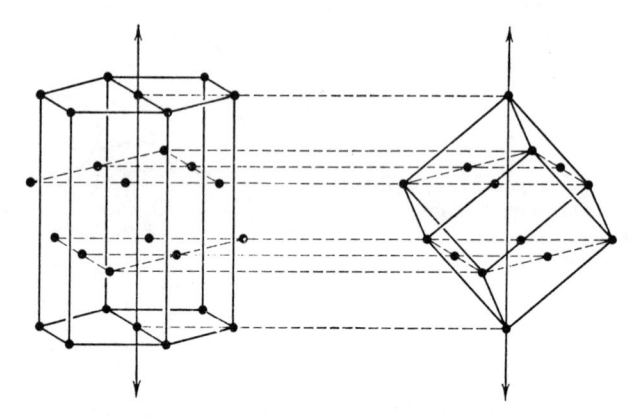

Fig. 140. Showing how the ABCA close-packed structure on the left can be represented as a face centred cubic structure on the right

four sets of close-packed layers (Fig. 141) at right-angles to each other. The hexagonal close-packed structure contains only one set of parallel close-packed layers.

This leads to important physical properties amongst the metals. The four sets of parallel layers in the cubic close-packed structure

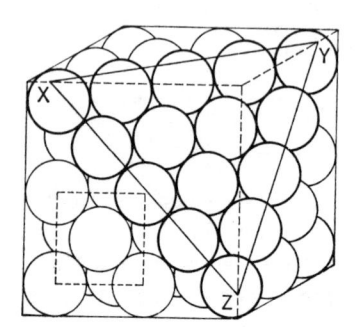

Fig. 141. A face centred structure, as in Fig. 140 showing the close-packing in the plane XYZ

means that metals with such a structure are ductile, malleable and rather soft. This is because there are many opportunities for gliding or slipping of one layer over another. Copper, silver, gold and nickel have the cubic close-packed structure. The opportunities for gliding or slipping are lessened in the hexagonal close-packed structure so that metals with this structure, e.g. chromium, vanadium, molybdenum and beryllium, are less malleable, harder and more brittle. Iron can adopt both structures, according to its heat treatment, and this is responsible for many of the special properties of iron.

Fig. 142. Cubic close-packed or face-centred cubic structure showing a co-ordination number of 12

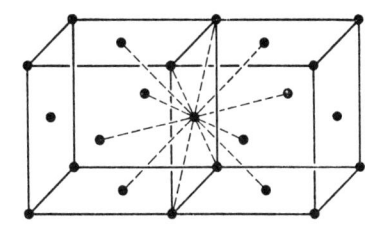

Figure 142 shows the most convenient way of representing the face-centred cubic structure, and also shows that the co-ordination number is 12.

12 Body-centred cubic structure The face-centred cubic structure, as shown in Fig. 142, has atoms at the corners of a cube and at each face centre. The body-centred cubic structure (Fig. 143a) has atoms at the corners of a cube and at the centre of the cube.

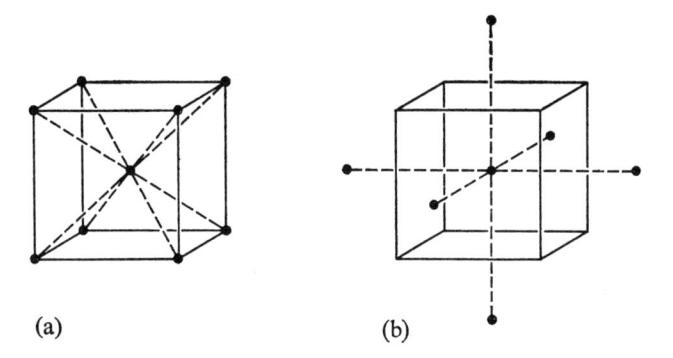

(a)   (b)

Fig. 143. Body-centred cubic structure showing the eight nearest neighbours at (a) and the six next-nearest neighbours at (b)

The body-centred structure is slightly less close-packed than the face-centred. In the former, the volume of space occupied by each sphere of radius r is $6 \cdot 6r^3$, whereas it is $5 \cdot 66r^3$ in the latter. In other words, the body-centred structure has more empty space in it.

The co-ordination number of the body-centred cubic structure is 8, the eight nearest neighbours being at a distance of $0·866r$. There are, however, six other fairly near neighbours, at a distance of r (Fig. 143b). As the distance between the nearest and next nearest neighbours is relatively small (15 per cent) the co-ordination number of the structure is sometimes given as 14 or as $(8+6)$.

13 Inter-relationship of simple crystal structures When spheres are packed together in a close-packed structure they only touch each other

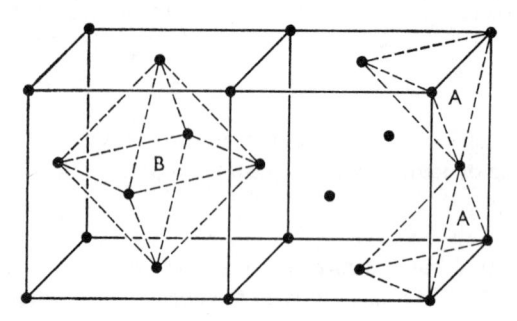

Fig. 144. Showing two tetrahedral sites, A, and one octahedral site, B, in a cubic close-packed structure

at certain points and there are holes between the spheres. Some of the holes may be described as tetrahedral holes; others as octahedral holes. Anything occupying a tetrahedral hole would be surrounded by four equidistant spheres arranged tetrahedrally; anything occupying an octahedral hole would be surrounded by six equidistant spheres arranged octahedrally (Figs. 144 and 145).

Simple geometrical calculations show that an octahedral hole is larger than a tetrahedral hole. For close-packing of spheres of radius r, a tetrahedral hole will accommodate a sphere of radius $0·225r$ whilst there is room in an octahedral hole for spheres of radius $0·414r$.

The holes are known, collectively, as *interstitial holes or sites* and when they are occupied the result is called an interstitial structure, an interstitial compound or a *solid solution*.

The size of the tetrahedral and octahedral holes in a crystal of a metal is too small to accommodate other metallic atoms, but smaller atoms such as those of hydrogen, boron, carbon and nitrogen do form interstitial structures. The outstanding example is provided by the interstitial structure of carbon in the face-centred cubic form of iron (γ-iron) which forms the basis of steel. The character of the various types of steels is dependent on the amount of carbon in the structure and its distribution.

In a truly close-packed structure the close-packed units are in contact with each other, but the same geometrical arrangement of centres can be maintained even though the units concerned do not quite touch each other, i.e. are pushed apart. In such structures, the size of the tetrahedral and octahedral holes will be larger than in those of truly close-packed structures. Common crystal structures can be

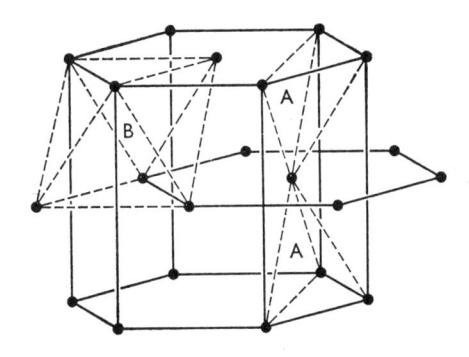

Fig. 145. Showing two tetrahedral sites, A, and one octahedral site, B, in a hexagonal close-packed structure

viewed, on this basis, as being derived from 'pseudo' close-packed structures in which all or some of the interstitial sites are occupied. The following summary shows the relationship between some common crystal structures:

Sites occupied	Cubic close-packed structure (*Fig. 144.*)	Hexagonal close-packed structure (*Fig. 145*)
All octahedral	Sodium chloride structure (Fig. 22)	Nickel arsenide structure
Half tetrahedral	Zinc-blende structure (Fig. 84)	Wurtzite structure (Fig. 85)
All tetrahedral	Fluorite structure (Fig. 24)	

The various figures of these related structures should be carefully compared.

Appendix

RADII AND ENERGY LEVELS OF STATIONARY STATES IN THE HYDROGEN ATOM

1 Bohr's treatment In order to calculate the radii and energies of the stationary states, Bohr made the arbitrary assumption that the angular momentum of an electron in any stationary state must be an integral-multiple of $h/2\pi$.

There is no *a priori* justification for this, but by using the assumption Bohr was able to obtain theoretical results in close agreement with experimental observations. The various simple calculations, using recommended SI units, are carried out as follows.

(*a*) *Radii of stationary states.* Angular momentum of electron of mass m kg, travelling with velocity v m s^{-1}, in a circular orbit of radius r m,

$$= mvr$$

Therefore on Bohr's assumption

$$mvr = \frac{nh}{2\pi} \qquad (a)$$

Attractive force (in newtons) between nucleus (charge, $+e$ C) and electron (charge, $-e$ C)

$$= \frac{e^2}{4\pi\varepsilon_0 r^2} \qquad (b)$$

where ε_0 is the permittivity of a vacuum.

Therefore potential energy of electron

$$= \int_\infty^r \frac{e^2}{4\pi\varepsilon_0 r^2} = -\frac{e^2}{4\pi\varepsilon_0 r}$$

(The negative sign indicates that work must be done to remove the electron away from the nucleus.)

Acceleration of electron towards centre of orbit

$$= \frac{v^2}{r}$$

Therefore, force on the electron

$$= \frac{mv^2}{r}$$

As the electron remains in its circular orbit, this force must be equal to the attractive force between the nucleus and the electron, therefore,

$$\frac{mv^2}{r} = \frac{e^2}{4\pi\varepsilon_0 r^2}$$

Kinetic energy of the electron

$$= \tfrac{1}{2}mv^2 = \frac{e^2}{8\pi\varepsilon_0 r} \qquad (c)$$

Total energy of electron = Potential energy + Kinetic energy

$$= -\frac{e^2}{4\pi\varepsilon_0 r} + \frac{e^2}{8\pi\varepsilon_0 r} = -\frac{e^2}{8\pi\varepsilon_0 r} \qquad (d)$$

From (a) and (c) it follows that

$$r = \frac{\varepsilon_0 n^2 h^2}{\pi m e^2}$$

and this expression gives the values of the radii of each stationary state, n having values 1, 2, 3 ... etc.

For $n=1$ and using the values $h = 6{\cdot}625\,6 \times 10^{-34}$ J s, $e = 1{\cdot}602\,1 \times 10^{-19}$ C, $m = 9{\cdot}109 \times 10^{-31}$ kg and $\varepsilon_0 = 8{\cdot}854 \times 10^{-12}$, the value of the radius obtained is $0{\cdot}529 \times 10^{-11}$ m.

(b) *Energy of stationary states.* Substituting the expression for r into the expression for the total energy of the electron given in (d) it follows that

$$\text{Total energy of electron} = -\frac{e^4 m}{8\varepsilon_0^2 n^2 h^2}$$

Using the numerical values given above this leads to an energy of $217{\cdot}9 \times 10^{-20}$ J when $n=1$, and $54{\cdot}48 \times 10^{-20}$ J when $n=2$ (see Fig. 4, page 14).

In general the change in energy when an electron passes from an orbit, $n = n_2$, to an orbit, $n = n_1$, is given by

$$\frac{e^4 m}{8\varepsilon_0^2 h^2}\left(\frac{1}{n_1^2} - \frac{1}{n_2^2}\right) \qquad (e)$$

This corresponds to radiation of wavelength λ given by

$$\frac{1}{\lambda} = \frac{e^4 m}{8c\varepsilon_0^2 h^3}\left(\frac{1}{n_1^2} - \frac{1}{n_2^2}\right)$$

This expression for the wavelength of the radiation causing a line in the hydrogen spectrum corresponds to the empirical expression relating the lines in the various spectral series,

$$\frac{1}{\lambda} = R_\text{H} \left(\frac{1}{n_1^2} - \frac{1}{n_2^2}\right)$$

given on page 8. It follows that R_H, known as the Rydberg constant, should be found equal to

$$\frac{e^4 m}{8 c \varepsilon_0^2 h^3}$$

if the Bohr treatment is accurate. This is, in fact, so.

(c) *Ionisation energy.* The expression given at (e) also enables the ionisation energy of hydrogen to be calculated. This represents (see page 14) the energy required to remove the electron in its normal state away from the atom, i.e. to infinity. Thus in the expression, (e), $n_2 = \infty$ and $n_1 = 1$, so that the ionisation energy is given by $e^4 m / 8\varepsilon_0^2 h^2$, which gives the value of 217.9×10^{-20} J, i.e. 13·71 electron-volt or 1322 kJ mol^{-1}, as is observed.

Formula Index

Ag$^+$, 39
Ag^{3+}, 39
AgBr, 58
AgCl, 58
AgI, 58, 133
Al, 42, 217
AlF$_3$, 42
Al$_2$Cl$_6$, 36
AlN, 133
AlP, 132
Ar, 28, 155
As, 221
AsH$_3$, 221
Au$^+$, 39
Au^{3+}, 39

B, 42, 107
BaF$_2$, 60
BaO, 58
BaS, 58
B(CH$_3$)$_3$, 100, 110
B(C$_6$H$_5$)$_3$, 110
BCl$_3$, 100, 107, 110
Be, 5, 107
BeCl$_2$, 100, 108
BeF$_4^{2-}$, 102
BeO, 133
BeS, 132
BF$_4^-$, 112
Bi, 41
Bi^{3+}, 40
BH$_4^-$, 112
BO$_3^{3-}$, 111
Br$_2$, 95, 156

C, 5, 107, 131, 156, 218, 219
Ca, 38
CaBr$_2$, 33
CaC$_2$, 58
CaCO$_3$, 58
CaF$_2$, 59, 60
CaO, 58
CaS, 58
CCl$_4$, 33, 112, 131
Cd^{2+}, 39
CdBr$_2$, 158
CdCl$_2$, 157
CdF$_2$, 60
CdI$_2$, 158
CdO, 158
Cd(OH)$_2$, 158
CdS, 132, 133, 158
(CF)$_n$, 157

≡CH, 140
CHCl$_3$, 112
CH$_2$Cl$_2$, 112
CH$_3$.COOH, 146
CH$_3$.NC, 35
CH$_3$.NO$_2$, 35
CH$_4$, 33, 101, 112, 135, 137
CH$_4 \cdot \frac{3}{4}$H$_2$O, 161
C$_2$H$_2$, 34, 109, 139
C$_2$H$_4$, 34, 139
C$_2$H$_4$(NH$_2$)$_2$, 163
C$_2$H$_5$OH, 138
C$_2$H$_5$SH, 138
C$_2$H$_6$, 139, 158
(CH$_3$)$_2$NH, 146
(CH$_3$)$_3$N, 146
C$_4$H$_6$, 115
[(CH$_3$)$_4$N]OH, 146
C$_6$H$_4$(NO$_2$)OH, 147
C$_6$H$_4$(OH)$_2$, 160
[C$_6$H$_4$(OH)$_2$]$_3$A, 161
C$_6$H$_6$, 115, 122
C$_{24}$HSO$_4$.2H$_2$SO$_4$, 157
Cl$_2$, 33, 95, 156
ClF$_3$, 102
Cl$_2$.8H$_2$O, 161
ClO$_4^-$, 112
Co^{2+}, 39
Co^{3+}, 39
CO, 98, 120
CO$^+$, 98
CO$_2$, 117, 119
CO$_3^{2-}$, 121
C$_{2.9}$O, 157
C$_{3.5}$O, 157
Co(CN)$_6^{3-}$, 175
Co$_2$(CO)$_6$.C$_2$H$_2$, 162
[Co(en)$_2$Cl$_2$]Cl, 178
CoF$_2$, 60
Co(NH$_3$)$_3$Cl$_3$, 162
Co(NH$_3$)$_6^{3+}$, 167, 175
Co(NH$_3$)$_4$Cl$_2$Cl, 177
Cr(C$_6$H$_6$)$_2$, 162
Cr(CN)$_6^{3-}$, 167
Cr(CO)$_6$, 162, 206
CrF$_2$, 197
Cr(H$_2$O)$_6^{3+}$, 167, 169
Cr(NH$_3$)$_6^{3+}$, 114, 167
Cs, 216
CsBr, 59
CsCl, 59
CsCN, 59
CsF, 57

CsI, 59
Cu, 219
Cu$^+$, 38, 39
Cu^{2+}, 39
CuBr, 132
CuCl, 132
CuCl$_2$, 197
Cu(CN)$_4^{2-}$, 190
Cu(CN)$_4^{3-}$, 175
Cu(en)$_2^{2+}$, 190
Cu(en)$_2$I$_2$, 174
CuF$_2$, 197
CuI, 132
CuI$_2$, 174
Cu(H$_2$O)$_4^{2+}$, 190
Cu(NH$_3$)$_6^+$, 114, 175
Cu(NH$_3$)$_4^{2+}$, 114, 168, 190
CuSO$_4$·H$_2$O, 146–7
CuSO$_4$·5H$_2$O, 146–7
CuSO$_4$·5NH$_3$, 147

D, 6
DNA, 151

EDTA, 164

F$_2$, 95, 156
Fe, 225
Fe^{2+}, 39
Fe^{3+}, 39
Fe(C$_6$H$_5$)$_2$, 162
Fe(CN)$_6^{3-}$, 114, 166, 176, 186
Fe(CN)$_6^{4-}$, 114, 166, 176
Fe(CO)$_5$, 206
FeF$_2$, 60
FeF$_6^{3-}$, 167, 186
Fe(H$_2$O)$_6^{2+}$, 176
Fe(H$_2$O)$_6^{3+}$, 176
Fe(NH$_3$)$_3^{2+}$, 167
Fe(phen)$_3^{2+}$, 176
Fe(phen)$_3^{3+}$, 176

Ga, 40, 221
Ge, 131, 219, 220–1

H, 5
H$^+$, 141
H$^-$, 141
H$_2$, 33, 91
H$_2^+$, 91, 142
HBr, 109
HCl, 98, 109, 126, 136
He, 5, 28, 135
He$_2$, 92
He$_2^+$, 91
HF, 101, 109, 143–4
HF$_2^-$, 142
Hg, 41
Hg^{2+}, 39
HgF$_2$, 42, 60
HgS, 132
HI, 109
H$_2$O, 101, 105, 110, 140, 143–4

H$_3$O$^+$, 102, 141
H$_2$S, 105
H$_2$Se, 105
H$_2$Te, 105

I$_2$, 95, 156
ICl$_2^-$, 102
ICl$_4^-$, 103
IF$_5$, 103
In, 221
In$^+$, 40
In^{3+}, 40

K, 216
K$^+$, 38
KBr, 57
KC$_8$, 157
KC$_{16}$, 157
KC$_{24}$, 157
KC$_{40}$, 157
KCl, 57
KF, 57
KHF$_2$, 142
KI, 57
K$_2$O, 60
K$_2$S, 33, 60
Kr, 28, 155

Li, 5, 209, 216
Li$_2$, 92
LiH, 42, 141
Li halides, 57
Li$_2$O, 60
Li$_2$S, 60

Mg, 217
MgF$_2$, 60
MgO, 58
MgS, 58
MnF$_2$, 60
MnO, 58
MnO$_2$, 60
Mn$_3$O$_4$, 194
MnS, 58
MoS$_2$, 158

N, 5
N^{3-}, 38
N$_2$, 93
Na, 209–11, 216, 217
NaCl, 32, 57–8, 131
Na$_2$O, 60
Na$_2$S, 60
Ne, 28, 155
NH$_2^-$, 102
NH$_3$, 101, 105, 110, 143–4
NH$_4^+$, 36, 102, 112
NH$_4$Br, 59, 147
NH$_4$Cl, 54, 59, 147
NH$_4$F, 147
NH$_4$F, 147
NH$_4$I, 59, 147
Ni^{2+}, 39

Ni^{3+}, 39
NiCl$_2$, 197
Ni(CN)$_4^{2-}$, 114, 168, 197
Ni(CN)$_2$.NH$_3$.C$_6$H$_6$, 161
Ni(CO)$_4$, 162, 206
Ni(H$_2$O)$_6^{2+}$, 171, 188
Ni(NH$_3$)$_6^{2+}$, 167, 171
NiO, 58
NO, 98, 120
NO$^+$, 96
N$_2$O, 120
—NO$_2$, 121
NO$_3^-$, 120

O$_2$, 94
O$_2^{2-}$, 128
OH$^-$, 102
O$_2$PtF$_6$, 36

P$_4$, 155
P^{3-}, 38
Pb^{2+}, 40
Pb^{4+}, 40
Pb(C$_2$H$_5$)$_4$, 112
PbCl$_4$, 40
PbO$_2$, 40, 60
PbS, 58
PCl$_5$, 102, 113
PdCl$_4^{2-}$, 168
PF$_6^-$, 113
PH$_3$, 106
PtCl$_4^{2-}$, 168
Pt(NH$_3$)$_2$Cl$_2$, 177

Rb, 216
Rb halides, 57
ReCl$_4^-$, 187
[Rh(SO$_2$N$_2$H$_2$)$_2$(H$_2$O)$_2$]$^-$Na$^+$, 178
Rn, 28

S$_8$, 154
Sb, 221
Sb^{3+}, 40
SbH$_3$, 106
Se, 155
SF$_6$, 103, 112
Si, 131
SiC, 132

SiCl$_4$, 112
SiF$_6^{2-}$, 113
SiH$_4$, 112
Sn, 131, 219
Sn^{2+}, 40
Sn^{4+}, 40
SnBr$_4$, 112
SnCl$_2$, 40, 100
SnCl$_4$, 112
SnF$_4$, 42
SnO, 40
SnO$_2$, 60
SO$_4^{2-}$, 112, 121
SrF$_2$, 60
SrO, 58
SrS, 58

T, 6
Te, 155
TeCl$_4$, 102
Ti(H$_2$O)$_6^{3+}$, 188, 201
TiO$_2$, 60
Tl^{2+}, 40
Tl^{4+}, 40
TlCN, 39
TlCl, 40, 59
TlBr, 40, 59
TlI, 40, 59
Tl$_2$SO$_4$, 40
Tl$_2$S, 40

U, 6

Xe, 36, 155
XeF$_2$, 36
XeF$_4$, 36
XeF$_6$, 36
Xe.6H$_2$O, 161
XeO$_3$, 37
XeOF$_4$, 36
XeO$_2$F$_2$, 6
XePtF$_6$, 36

Zn^{2+}, 38, 39
ZnF$_2$, 60
Zn(NH$_3$)$_4^{2+}$, 114, 167
ZnO, 133
ZnS, 131–2

233

Subject Index

Abegg, 22
Absorption spectra, 8
Acceptor, 35
Acceptor ligand orbital, 204
Acetyl-acetone, 163
Actinides, 22
Actinons, 22
Adenine, 151
Alanine, 149
Alcohols, 145
Alkali metal hydrides, 141
α-helix, 150
a metals, 173
Amines, 146
Amino acids, 149
Angström unit, 7
Angular dependence, 76
Angular distribution of bonds, 76–9
Anion formation, 38
Anti-bonding orbitals, 83, 200
Anti-fluorite structure, 60
Appendix, 228
Arithmetic mean, 124
Aseptic distillation, 148
Association, 143
Atomic crystals, 131
Atomic orbital, 69
Atomic orbitals, combinations of, 85–6
Atomic spectra, 8
Atomic structure, 4
Atomic volume, 20, 42
Aufbau principle, 26, 89
Avogadro's hypothesis, 2
Azimuthal quantum number, 16

Back bonding, 204
Back donation, 204
Balmer, 8
Band model, 209
Band spectra, 212
Bartlett, 36
Base pairing, 151
Basolo, 176
Benzene, 115–16, 122, 158
Benzene-1,4-diol, 160
Berzelius, 1
Bethe, 180
2,2'-bipyridine, 163
Bidentate group, 163
Binding energy in metals, 216
Black body radiation, 10
b metals, 173
Body-centred cubic structure, 221, 225

Bohr, 10, 11, 23
Bohr magneton, 64
Boiling points, abnormal, 144
Bond-dissociation energy, 135, 140
Bond energies, table of, 137
Bond energy, average, 135
Bond energy of hybrid bonds, 139
Bonding orbitals, 83, 200
Bond order, 96
Born-Haber cycle, 62–3
Boundary surface, 71
Brackett, 8
B sub-group elements, 38
Buta-1,3-diene, 115

Caesium chloride crystal structure, 58–9
Calcite, 58
Canonical form, 118
Carbonyls, 162
Carboxylic acids, 145
Cartesian coordinates, 70
Cation formation, 38–9
Cell diagram, 91
Charge cloud, 73
Chelate groups, 163
Chelation, 147, 163, 174
Chlorides, fused, 43–4
Chlorine, compounds of, 129
Chromium, compounds of, 129, 191
Clathrates, 160–1
Cobalt(III) complexes, 175, 191
Co-ionic bond, 34
Colour of complexes, 190
Complex anions, 164
Complex cations, 164
Complex compounds, 162–79
Complex compounds, stability of, 169–177
Conductivity, 217–19
Conductors, 217
Conjugated system, 116
Continuous spectra, 8
Coordinate bond, 35, 165
Coordination compounds, 162–79
Coordination number, 55, 162
Coordination number and ionic radii, 54
Copper(I) complexes, 175
Coulomb's law, 61
Covalency maxima, 113
Covalent bond, 33

Covalent bond, directed, 100–14
Covalent bond, energy changes in formation, 135
Covalent bond in crystals, 131–2
Covalent bond, ionic character, 122
Covalent crystals, 131
Covalent radii, 133–5
Crystal energy, 61–2
Crystal field splitting, 181
Crystal field stabilisation energy, 191–5
Crystal field theory, 180–90
Crystal field theory, applications of, 190–7
Crystal structures, covalent, 131
Crystal structures, interrelationships, 226–7
Crystal structures, ionic, 55–61
Crystal structures, metallic, 221
Crystal structures, molecular, 154
Cubic close-packed structures, 221, 223
Cytosine, 151

Dative bond, 34
Davisson, 68
d-block elements, 31
de Broglie, 68, 79
Debye, 159
Debye unit, 126
Degeneracy, 18
Degeneracy of *d*-orbitals, 180
δ-orbitals, 87
Δ, 182, 187–9
Deoxythymidine-5′-phosphate, 151
Deposition pressure, 49
d_ε-orbitals, 78
Deuterium, 6
d_γ-orbitals, 78
Diagonal relationship, 43
Diamagnetism, 64–5
Diamond structure, 131
Diffraction, 68
Diffuse series, 9
Digonal hybridisation, 108
Dimers, 145
Dimethyl glyoxime, 163
Dipole moment, 126, 158–9
Directed covalent bonds, 100–14
Dispersion effect, 159
Disproportionation, 175
Dissociation energy, 96–7
DNA, 151–3
Dobereiner, 20
Donor, 35
Donor ligand orbitals, 204
d-orbital, 78
Double-bond covalent radii, 134
Dq, 182, 187
dsp^2, hybridisation, 168
d^2sp^3 hybridisation, 112, 166
Dualistic theory, 1
Dumas, 2

EDTA, 164
Eigenvalues, 69
18-electron group, 38
e_g-electrons, 195
e_g-orbitals, 78
Einstein, 11
Electrochemical series, 50
Electrode potentials, 49–50
Electrolytic solution pressure, 49
Electromagnetic spectra, 7
Electron, 5
Electron affinity, 48, 125
Electronegative elements, 33
Electronegativity values, 125
Electron spin, 16
Electron volt, 14
Electrons, arrangement in orbitals, back end paper, 24–28
Electrons, promotion of, 106
Electron-pair distribution, 100–3
Electropositive elements, 33
Elements, type of, 28–31
Emission spectra, 8
Ethylene diamine, 163
Ethyl 3-oxobutanoate, 148
Excited state, 107
Exclusion principle, 16

Face-centred cubic structure, 221, 223
Fajans' rules, 42
f-block elements, 31
Ferroin, 176
Ferromagnetism, 64
Fluorite structure, 59
f-orbitals, 78
Frankland, 3
Friedrich, 23
Fundamental series, 9

Gas equation, 154
Geometric isomerism, 177
Geometric mean, 124
Gerhardt, 2
Germer, 68
Giant molecules, 131
Goudsmit, 16
Graphite, 156
Graphitic compounds, 157
Ground state, 13
Group theory, 198
Guanine, 1, 51

Halogens, crystal structures of, 156
Heat of atomisation, 136
Heat of formation, 169
Heat of hydration, 169, 191
Heisenberg, 79
Helium molecule, 92
Helium molecule ion, 91
Heteronuclear diatomic molecules, 96

Hexadentate groups, 164
Hexacyanoferrate(III) ion, 166, 186
Hexacyanoferrate(II) ion, 167, 186
Hexafluoroferrate(III) ion, 167, 186
Hexagonal close-packed structures, 221, 223
High spin complexes, 186
Homologous series, 3
Homonuclear diatomic molecules, 91–6
Huyghens, 67
Hybrid bonds, 106–14, 165–9
Hybridisation of bonds, 106–14, 165–9
Hybrid orbitals, summary, 114
Hydrated transitional ions, 169, 191
Hydrides, association of, 143–4
Hydrides, melting and boiling points of, 143–4
Hydrides, molecular, 142
Hydrides of alkali metals, 141
Hydrides, salt-like, 141
Hydrides, volatile, 142
Hydrocarbon complexes, 162, 207
Hydrogen bond, 141–9
Hydrogen, isotopes of, 6
Hydrogen molecule, 91
Hydrogen molecule ion, 91
Hydrogen spectrum, 9
Hydrogen, valency of, 141
Hydronium ion, 141
Hydroxonium ion, 141

Ice, 144–5
Iceland spar, 58
Inclusion compounds, 160
Induction effects, 159
Inductive effect, 123
Inert complexes, 176
Inert pair effect, 39
Ingold, 118
Inner-orbital complex, 166
Inorganic isomerism, 177–9
Instability constant, 171
Insulators, 217
Interstitial holes, 226
Intrinsic semi-conductor, 219
Inverted spinel, 194
Ionic bond, 32, 38–50
Ionic character of covalent bond, 122–3, 126–7
Ionic compounds, characteristics of, 51
Ionic crystals, 55–61
Ionic radii, 51–5
Ionic resonance energy, 124
Ionic structures, 38–9
Ionisation energy, 44, 125
Ionisation energy, table of, 46
Ionisation isomerism, 179
Ionisation potential, 44
Ion pair, 32, 58
Ions, arrangement in crystals, 55–61

Ions, ease of formation, 42
Ions, limitations to formation, 41
Ions, magnetic moments, 63–5
Iron complexes, 176
Iso-electronic principle, 96, 102
Isomerism, 177–9
Isotopes, 6

Jahn-Teller, effect, 195–7

Kekulé, 3
Keesom, 158
Kossel, 32

Labile complexes, 176
Lanthanide contraction, 54
Lanthanides, 22
Lanthanons, 22
Lattice energy, 61–2
Laue, 23
Law of octaves, 20
Layer lattices, 157
LCAO approximation, 81
Lewis, 33
Ligand field theory, 198–209
Ligands, 162
Linear hybridisation, 108
Line spectra, 8
Lithium molecule, 92
Localised orbitals, 114
London, 159
Lone pair, 35, 101–3
Lothar Meyer, 20
Low-spin complexes, 186
Lummer, 10
Lyman, 8, 15

Magnetic moment of ions, 63–5, 165–9
Magnetic quantum number, 17
Manganese, compounds of, 129
Maximum multiplicity, rule of, 27
Maximum overlap, 104
Melting points, abnormal, 144
Mendeleef, 20
Mercaptans, 145
Mesomerism, 118
Metallic bond, 209–18
Metallic crystals, 221
Methionine, 149
Meyer, 148
Molecular crystals, 131, 154
Molecular hydrides, 142
Molecular orbitals, 81–90
Molyslip, 158
Monodentate ligands, 163
Moore, 146
Moseley, 15, 24
Mulliken, 90, 124
Multiple bond covalent radii, 134

Natural order of stability, 173
$n(E)$ curves, 213
$N(E)$ curves, 215

Neutron, 5
Newlands, 20
Newton, 67
Nickel(II) complexes, 168, 193
Nitrogen, compounds of, 129
Nitrogen molecule, 92–3
Nitrogen oxide, 98
Nitrogen oxide, paramagnetism, 98
Nitromethane, 35
Nitrosyl ion, 96
Noble gas compounds, 36
Noble gases, 22, 28, 155, 159
Noble gas structures, 28, 38
Nodal sphere, 72
Nodal surface, 72
Normal metals, 222
Normal state, 13
N-type semi-conductors, 220
Nucleic acids, 151–3
Nucleotides, 151–3

Octahedral complexes, 166
Octahedral field, 182
Octahedral hybridisation, 112
Optical isomerism, 177–8
Orbitals, combination of, 86, 198
Orbitals, delocalised, 114
Orbitals, hybrid, 107, 114
Orgel, 180
Orientation effect, 158
o-substitution, 147
Outer-orbital complexes, 166
Overlap of orbitals, 104
Oxidation, 129
Oxidation number, 127
Oxidation state, 127
Oxides, bond distances in, 195

Paramagnetism, 63–5, 94, 98
Paschen, 8
Pauling, 52, 117, 124, 165, 180
Pauli principle, 16
p-block elements, 30
Pearson, 176
Peptide link, 149
Periodic table, anomalous placings, 22
Periodic table, front end-paper, 20–4, 29
Periods, 21
Peroxides, 128
Pfund, 8
Phenanthroline, 163
Photo-electric effect, 67
Photon, 67
π-bonding in complexes, 202–9
π-orbitals, 84
Planck, 10
Planck's constant, 11
Plastic sulphur, 155
Pleated sheets, 151
Polar coordinates, 70
Polymorphism, 221

Polypeptides, 149
p-orbitals, 96–8
Primary structure, 150
Principal quantum number, 15
Principal series, 9
Pringsheim, 10
Probability distribution, 73
Promotion of electrons, 106
Proteins, 149–51
Protium, 6
Proton, 5
Proton bonding, 143
P-type semi-conductors, 220

Quantum mechanics, 68
Quantum numbers, 15–19
Quantum theory, 10
Quantum, value of, 11
Quaternary base, 146

Radial probability distribution, 75
Radius ratio, 55
Rare earths, 22
Rare gases, 37
Reactivity series, 50
Reduction, 129
Resolving of orbitals, 181
Resonance, conception of, 117
Resonance, conditions for, 119
Resonance energy, 118
Resonance, example of, 119–22
Resonance hybrid, 118
Resonance, ionic, 124
Resonating molecules, 119–22
Rule of eight, 22
Rutherford, 11
Rutile, 59–60
Rydberg constant, 9

Sandwich molecules, 208
s-block elements, 30
Schomaker, 135
Schrödinger, 68
Schweitzer's solution, 168
Secondary structure, 150
Selection rules, 16
Semi-conductors, 219
Semi-polar double bond, 35
Shared pair, 33
Sharp series, 9
Shell of orbits, 16
Sidgwick, 164
σ-bond, 83
σ-orbital, 83
Single-bond covalent radii, 133
Sodium atoms, energy changes in, 17
Sodium chloride structure, 57–8
Soft X-ray spectra, 212
Solid solutions, 226
s-orbitals, 71–6
Spectra, 7–10
Spectra, absorption, 8, 188

Spectra, atomic, 8
Spectra, band, 8
Spectra, continuous, 8
Spectra, emission, 8
Spectra, line, 8
Spectra, hydrogen, 9
Spectra, X-ray, 23
sp-hybridisation, 108
sp^2 hybridisation, 110
sp^3 hybridisation, 112, 166
sp^3d^2 hybridisation, 112, 166
sp^3d hybridisation, 113
Spectral series, 8
Spectral terms, 9
Spectrochemical series, 189, 204
Spherically symmetrical, 72
Spinels, 193
Spin quantum number, 16
Splitting of orbitals, 181
Square planar complexes, 168, 193, 197
Square planar field, 183
Stabilisation of valency states, 174
Stability constant, 171
Stability, natural order of, 173
Standard electrode potential, 49
Stark effect, 15
Stationary states, 12, 228–30
Stevenson, 135
Stock system, 129, 164
Sub-group elements, 21
Subsidiary quantum number, 15
Sulphur crystal, 154

Taube, 176
Tautomerism, 118
Tertiary structure, 150
Tetrahedral complexes, 167
Tetrahedral field, 183
Tetrahedral hybridisation, 112
Tetramminecopper(II) ion, 168, 190
Tetracyanonickelate(II), 168
Thiophene, 161
Thomsen, 23
Thymine, 151
Transitional ions, 39, 65
Transition cations, 39, 65
Transition elements, 22, 31
Triamine propane, 163

Tridentate groups, 163
Triethylene tetramine, 174
Trigonal bi-pyramidal hybridisation, 113
Trigonal hybridisation, 110
Triple-bond covalent radii, 134
Tripyridine, 163
Tritium, 6
True metals, 222
t_{2g} orbitals, 78
Types, theory of, 2
Typical elements, 21
Tyrosine, 149

Uhlenbech, 16
Uncertainty principle, 78
Units, 13–15
Uranium, isotopes of, 6

Valency, as a number, 3, 127
Valency bonds, 32–6
Valency, definition of, 4
Valency electrons, 30
Valency, importance of, 1
van der Waals' forces, 154–61
van der Waals' radii, 159–61
van Vleck, 180
Volatile hydrides, 142
Voltaic pile, 1

Water, association of, 144
Water, bond angle, 106
Wave function, 68
Wave mechanics, 68
Wave nature of electron, 67–8
Wave number, 14
Werner, 162
Williamson, 2
Winmill, 146
Wurtzite, structure of, 131

X-ray spectra, 23, 212
Xenon, fluorides of, 36
Xenon, oxides of, 37

Zeeman effect, 15
Zinc blende structure, 131

Periodic Table

Period	Group 0	Group 1 A	Group 1 B	Group 2 A	Group 2 B	Group 3 A	Group 3 B	Group 4 A	Group 4 B	Group 5 A	Group 5 B	Group 6 A	Group 6 B	Group 7 A	Group 7 B	Group 8
1		1 H 1.007 97														
2 (1st short period)	2 He 4.0026	3 Li 6.939		4 Be 9.0122		5 B 10.811		6 C 12.011 15		7 N 14.006 7		8 O 15.999 4		9 F 18.998 4		
3 (2nd short period)	10 Ne 20.179	11 Na 22.989 8		12 Mg 24.312		13 Al 26.9815		14 Si 28.086		15 P 30.973 8		16 S 32.064		17 Cl 35.453		
4 (1st long period)	18 Ar 39.948	19 K 39.102	29 Cu 63.546	20 Ca 40.08	30 Zn 65.37	21 Sc 44.956	31 Ga 69.72	22 Ti 47.90	32 Ge 72.59	23 V 50.942	33 As 74.921 6	24 Cr 51.996	34 Se 78.96	25 Mn 54.938	35 Br 79.904	26 Fe 55.847 27 Co 58.933 2 28 Ni 58.71
5 (2nd long period)	36 Kr 83.80	37 Rb 85.47	47 Ag 107.868	38 Sr 87.62	48 Cd 112.40	39 Y 88.905	49 In 114.82	40 Zr 91.22	50 Sn 118.69	41 Nb 92.906	51 Sb 121.75	42 Mo 95.94	52 Te 127.60	43 Tc (99)	53 I 126.904 4	44 Ru 101.07 45 Rh 102.905 46 Pd 106.4
6 (3rd long period)	54 Xe 131.30	55 Cs 132.905	79 Au 196.967	56 Ba 137.34	80 Hg 200.59	57 La 138.91 58–71 The Rare Earths*	81 Tl 204.37	72 Hf 178.49	82 Pb 207.19	73 Ta 180.948	83 Bi 208.98	74 W 183.85	84 Po (210)	75 Re 186.2	85 At (210)	76 Os 190.2 77 Ir 192.2 78 Pt 195.09
7	86 Rn (222)	87 Fr (223)		88 Ra (226)		89 Ac (227) 90–103 The Actinons†										

Typical elements

* The Rare Earths or Lanthanons

58 Ce 140.12	59 Pr 140.907	60 Nd 144.24	61 Pm (147)	62 Sm 150.35	63 Eu 151.96	64 Gd 157.25	65 Tb 158.924	66 Dy 162.50	67 Ho 164.930	68 Er 167.26	69 Tm 168.934	70 Yb 173.04	71 Lu 174.97

† The Actinons

90 Th 232.038	91 Pa (231)	92 U 238.03	93 Np (237)	94 Pu (242)	95 Am (243)	96 Cm (247)	97 Bk (247)	98 Cf (249)	99 Es (254)	100 Fm (253)	101 Md (256)	102 No (253)	103 Lw (257)

THE PERIODIC TABLE (after Thomsen and Bohr)

1A	2A	3A	4A	5A	6A	7A	8			1B	2B	3B	4B	5B	6B	7B	0
1 H																	2 He
3 Li	4 Be											5 B	6 C	7 N	8 O	9 F	10 Ne
11 Na	12 Mg		←—— Transitional Elements ——→									13 Al	14 Si	15 P	16 S	17 Cl	18 Ar
19 K	20 Ca	21 Sc	22 Ti	23 V	24 Cr	25 Mn	26 Fe	27 Co	28 Ni	29 Cu	30 Zn	31 Ga	32 Ge	33 As	34 Se	35 Br	36 Kr
37 Rb	38 Sr	39 Y	40 Zr	41 Nb	42 Mo	43 Tc	44 Ru	45 Rh	46 Pd	47 Ag	48 Cd	49 In	50 Sn	51 Sb	52 Te	53 I	54 Xe
55 Cs	56 Ba	57* La	72 Hf	73 Ta	74 W	75 Re	76 Os	77 Ir	78 Pt	79 Au	80 Hg	81 Tl	82 Pb	83 Bi	84 Po	85 At	86 Rn
87 Fr	88 Ra	89† Ac															

* The Rare Earths or Lanthanons

58 Ce	59 Pr	60 Nd	61 Pm	62 Sm	63 Eu	64 Gd	65 Tb	66 Dy	67 Ho	68 Er	69 Tm	70 Yb	71 Lu

† The Actinons

90 Th	91 Pa	92 U	93 Np	94 Pu	95 Am	96 Cm	97 Bk	98 Cf	99 Es	100 Fm	101 Md	102 No	103 Lw

THE ARRANGEMENT OF ELECTRONS IN THE

	1s	2s	2p	3s	3p	3d	4s	4p	4d	4f	5s	5p	5d	5f	6s	6p	6d	7s	
1 H	1																		
2 He	2																		1s Full
3 Li	2	1																	
4 Be	2	2																	
5 B	2	2	1																
6 C	2	2	2																2s Full
7 N	2	2	3																
8 O	2	2	4																
9 F	2	2	5																
10 Ne	2	2	6																2p Full
11 Na	2	8		1															
12 Mg	2	8		2															
13 Al	2	8		2	1														3s Full
14 Si	2	8		2	2														
15 P	2	8		2	3														
16 S	2	8		2	4														
17 Cl	2	8		2	5														
18 A	2	8		2	6														3p Full
19 K	2	8		8			1												
20 Ca	2	8		8			2												4s Full
21 Sc	2	8		8		1	2												
22 Ti	2	8		8		2	2												
23 V	2	8		8		3	2												
24 Cr	2	8		8		5	1												
25 Mn	2	8		8		5	2												
26 Fe	2	8		8		6	2												
27 Co	2	8		8		7	2												
28 Ni	2	8		8		8	2												
29 Cu	2	8		8		10	1												
30 Zn	2	8		8		10	2												3d Full
31 Ga	2	8		18			2	1											
32 Ge	2	8		18			2	2											
33 As	2	8		18			2	3											
34 Se	2	8		18			2	4											
35 Br	2	8		18			2	5											
36 Kr	2	8		18			2	6											4p Full
37 Rb	2	8		18			8				1								
38 Sr	2	8		18			8				2								5s Full
39 Y	2	8		18			8		1		2								
40 Zr	2	8		18			8		2		2								
41 Nb	2	8		18			8		4		1								
42 Mo	2	8		18			8		5		1								
43 Tc	2	8		18			8		6		1								
44 Ru	2	8		18			8		7		1								
45 Rh	2	8		18			8		8		1								
46 Pd	2	8		18			8		10										
47 Ag	2	8		18			8		10		1								
48 Cd	2	8		18			8		10		1								4d Full
49 In	2	8		18			18				2	1							
50 Sn	2	8		18			18				2	2							

Transition elements: 21 Sc – 30 Zn

Transition elements: 39 Y – 48 Cd

ATOMS OF THE ELEMENTS IN THEIR NORMAL STATES

		1s	2s 2p	3s 3p 3d	4s 4p 4d 4f	5s 5p 5d 5f	6s 6p 6d	7s	
	51 Sb	2	8	18	18	2 3			
	52 Te	2	8	18	18	2 4			
	53 I	2	8	18	18	2 5			
	54 Xe	2	8	18	18	2 6			5p Full
	55 Cs	2	8	18	18	8	1		
	56 Ba	2	8	18	18	8	2		6s Full
	57 La	2	8	18	18	8 1	2		
	58 Ce	2	8	18	18 2	8	2		
	59 Pr	2	8	18	18 3	8	2		
	60 Nd	2	8	18	18 4	8	2		
	61 Pm	2	8	18	18 5	8	2		
	62 Sm	2	8	18	18 6	8	2		
	63 Eu	2	8	18	18 7	8	2		
	64 Gd	2	8	18	18 7	8 1	2		
	65 Tb	2	8	18	18 9	8	2		
	66 Dy	2	8	18	18 10	8	2		
	67 Ho	2	8	18	18 11	8	2		
	68 Er	2	8	18	18 12	8	2		
	69 Tm	2	8	18	18 13	8	2		
	70 Yb	2	8	18	18 14	8	2		
	71 Lu	2	8	18	18 14	8 1	2		4f Full
	72 Hf	2	8	18	32	8 2	2		
	73 Ta	2	8	18	32	8 3	2		
	74 W	2	8	18	32	8 4	2		
	75 Re	2	8	18	32	8 5	2		
	76 Os	2	8	18	32	8 6	2		
	77 Ir	2	8	18	32	8 7	2		
	78 Pt	2	8	18	32	8 9	1		
	79 Au	2	8	18	32	8 10	1		
	80 Hg	2	8	18	32	8 10	2		5d Full
	81 Tl	2	8	18	32	18	2 1		
	82 Pb	2	8	18	32	18	2 2		
	83 Bi	2	8	18	32	18	2 3		
	84 Po	2	8	18	32	18	2 4		
	85 At	2	8	18	32	18	2 5		
	86 Rn	2	8	18	32	18	2 6		6p Full
	87 Fr	2	8	18	32	18	8	1	
	88 Ra	2	8	18	32	18	8	2	7s Full
	89 Ac	2	8	18	32	18	8 1	2	
	90 Th	2	8	18	32	18	8 2	2	
	91 Pa	2	8	18	32	18 2	8 1	2	
	92 U	2	8	18	32	18 3	8 1	2	
	93 Np	2	8	18	32	18 5	8	2	
	94 Pu	2	8	18	32	18 6	8	2	
	95 Am	2	8	18	32	18 7	8	2	
	96 Cm	2	8	18	32	18 7	8 1	2	
	97 Bk	2	8	18	32	18 9	8	2	
	98 Cf	2	8	18	32	18 10	8	2	
	99 Es	2	8	18	32	18 11	8	2	
	100 Fm	2	8	18	32	18 12	8	2	
	101 Md	2	8	18	32	18 13	8	2	
	102 No	2	8	18	32	18 14	8	2	

Transition elements
Rare earths or lanthanons
Actinons